DISPUTE AVOIDANCE and RESOLUTION for CONSULTING ENGINEERS

Richard K. Allen

Published by
ASCE Press
American Society of Civil Engineers
345 East 47th Street
New York, New York 10017-2398

ABSTRACT

In his book, *Dispute Avoidance and Resolution for Consulting Engineers,* Richard K. Allen points out that the vast majority of disputes involving engineers arise from failures to adhere to the fundamentals of good engineering and business practices. Since few disputes come from the misunderstanding or misapplication of scientific principles, Mr. Allen addresses the larger framework of the practice of engineering by discussing the design and construction administration services provided by consulting engineers. He focuses on the fundamentals of a consulting engineer's practice and ways to avoid and resolve disputes. In order to do this, Mr. Allen explores such topics as: 1) Quality control; 2) service performance; 3) dispute analysis; 4) adversarial dispute resolution; 5) negotiation; and 6) alternative dispute resolution. Finally, Richard Allen reminds the reader that while the best dispute is the avoided dispute, when a dispute is unavoidable, the engineer should consider realistic settlement that focuses on the business realities of the situation.

Library of Congress Cataloging-in-Publication Data

Allen, Richard K., 1955-
Dispute avoidance and resolution for consulting engineers/Richard K. Allen, Esq.
p.cm.
Includes index.
ISBN 0-87262-903-1
1.Civil engineering contracts—United States. 2.Consulting engineers—Legal status, laws, etc.—United States. 3.Dispute resolution(Law)—United States. I.Title.
KF902.A743 1993 93-43314
343.73'078624—dc20 CIP
[347.30378624]

The material presented in this publication has been prepared in accordance with generally recognized engineering principles and practices, and is for general information only. This information should not be used without first securing competent advice with respect to its suitability for any general or specific application.

The contents of this publication are not intended to be and should not be construed to be a standard of the American Society of Civil Engineers (ASCE) and are not intended for use as a reference in purchase specifications, contracts, regulations, statutes, or any other legal document.

No reference made in this publication to any specific method, product, process or service constitutes or implies an endorsement, recommendation, or warranty thereof by ASCE.

ASCE makes no representation or warranty of any kind, whether express or implied, concerning the accuracy, completeness, suitability or utility of any information, apparatus, product, or process discussed in this publication, and assumes no liability therefor.

Anyone utilizing this information assumes all liability arising from such use, including but not limited to infringement of any patent or patents.

Library of Congress Catalog Card No: 93-43314
ISBN 0-87262-903-1
Manufactured in the United States of America.

Acknowledgments

The author gratefully acknowledges the support, assistance and encouragement of Elizabeth Fronduto Allen during the preparation of this manual.

The author acknowledges the support of the law firm of Gadsby & Hannah and, in particular, the firm's construction group which dedicates its practice to engineering and construction law. Special thanks is given to Gadsby & Hannah's support staff for assistance in processing the numerous revisions to this manuscript.

The author thanks the following people for their comments and assistance in the preparation of the manuscript: Gary D. Bates of the Roenker-Bates Group, Terry McManus at Intel, Frank Muller of Metro Mediation Services, Robert A. Rubin at Postner & Rubin Attorneys at Law, Dr. Jeffrey Russell of the University of Wisconsin-Madison, and Robert J. Smith at Wickwire Gavin.

Many of the specification sections and contract provisions cited herein were modeled after those used by Dufresne-Henry, Inc. Dufresne-Henry Inc.'s engineering practice is an example of responsible engineering that promotes dispute avoidance and resolution.

Finally, the author thanks Zoe G. Foundotos, Acquisitions Editor of the ASCE, for her thoughtful review and comments on the drafts of this manuscript, as well as her overall encouragement in this endeavor.

Foreword

After years of working as a consulting engineer and later as a lawyer who has both defended and prosecuted claims against engineers, there is no question that the vast majority of disputes involving engineers arise from failures to adhere to the fundamentals of good engineering and business practices. Few disputes arise from the misunderstanding or misapplication of scientific principles. For example, a failure to clearly articulate a specification requirement or coordinate drawings are much more common errors or omissions than design miscalculations. Yet, most engineers' focus is on technical competence and not practical competence.

Engineering schools at both the undergraduate and graduate levels exclusively teach scientific principles and their application, or in other words, technical competence. Technical competence alone is not nearly enough. A competent engineer must be able to draw on his or her technical competence within the framework of the practice of engineering. The practice of engineering demands written and oral communication skills and an understanding of government, financing, regulations, ethics, business, and law. Some of the practice of engineering can and should be taught in the classroom; most is learned through experience.

In my legal practice, I find it truly remarkable the number of highly trained and skilled engineers from large and small firms alike who simply do not appreciate the consulting engineer's role, responsibilities, and limitations. These lessons are too often learned the hard way in the context of disputes and claims, none of which are pleasant, some of which are disastrous.

This book addresses the practice of engineering in its most common manifestation, the design and construction administration services provided by consulting engineering firms. It focuses on the fundamentals of a consulting engineer's practice, the disputes that arise from this practice, and ways to avoid and resolve these disputes.

This manual is not legal advice, nor is it written for lawyers, despite the fact that it is written by a lawyer (who is also a civil engineer). There are few case citations, little reference to statutes, and no footnotes. I hope my colleagues who are attorneys are not disappointed in what would otherwise be an unforgivable omission in a legal publication.

This manual is designed as a basic reference of the fundamentals and real-world practice and problems of a consulting engineer. To engineering students, I invite you to discover the practice of engineering. To experienced designers and construction project administrators, I invite you to revisit much of what you have already learned, or will learn, the hard way.

Boston,

R. K. A.
May 1993

Contents

Table of Figures

I.

Introduction

What better introduction to the consulting engineering practice and the avoidance of disputes than a case study of malpractice creating disputes. A hypothetical, and notably exaggerated, history of an engineer's work on a project gone bad through the eyes of a seasoned professional could read as follows:

1.0 The Hypothetical

After four years of undergraduate work and possibly two additional years of post-graduate study, you have the right to expect the junior engineer you just hired to know what she is doing. Put her on a project team, get her productive, and get something back on that salary you are shelling out. It probably makes the most sense to get her started on a big budget project with lots of team members. In that way, you can bury any inefficiency in the bills; the client will never know the difference.

Immediately assigning this junior engineer to project work gives you time to do just a bit more marketing, something that is taking up most of your time these days anyway. Marketing has to be done, so why not sell the firm with your extensive experience and knowledge that unfortunately has not seen the light of any real project work in years. That is the price of success. You must keep bringing home the projects. Otherwise, junior engineers, like the one you just hired, could not be employed.

Ah, the budget. You have not thought about the budget in a while. Thank goodness the project is cost plus and you are only 75% expended and 25% complete. You have 25% more budget to spend while figuring out how to explain to the client that you need 50% more. You had to offer a very tight, and probably unrealistic, price proposal in the first place, or the client would have hired a competitor. Despite the fact that you know your firm to be better qualified, this client (like most clients) looked at the bottom line when it came to hiring a consultant.

Your statement of qualifications was probably ignored when you responded to the request for proposals with a "budget price." You certainly could not explain at the short list presentation that if the client wants the best, it costs more. Instead, you priced the project to beat the competition, and promised the potential client that your services would be both extensive and of the highest quality. Otherwise, you would have lost this project and every other project you were forced to bid.

You simply cannot tell the client you underpriced the project. Hence, a different approach is necessary. One of the more successful alternatives is to claim that the project scope has changed. You are now required to do more than you originally agreed to perform. This will work because the client does not really understand what your firm is doing anyway and the scope of services is simply a somewhat vaguely written letter agreement that can be creatively interpreted. If the client gets somewhat upset at the increase in cost, you can always try to recover any loss during construction services. Keep in mind that you are in the driver's seat — no client will fire you in the middle of a project design. In any case, you would have negotiated a fair budget if not for the fact that the client required you to bid your services in the first place.

If you still need to trim the budget to avoid an overrun, even after a contract increase, you can always cut back on services. Again, creative interpretation of the scope of services comes to the rescue. Let's cut back on quality control, it's a moving target and can mean anything from a cursory look at the plans and specifications to an independent, detailed, time-consuming review. The cursory look is surely cheaper and will keep your cost down. Your client will appreciate more expedient quality control because you're running out of time to complete the design and extensive quality control will only make the project late. Timeliness is important.

Well, you have done it. The project is now 90% complete, the budget has been increased by 50%, and 100% of the increased budget has been expended. That last 10% of the design can be picked up in addenda or change orders during construction under a new budget so there is no need to worry. The client has now approved the plans and specifications, not having a clue as to their overall quality. In no time at all you and the client are ready to move forward into the construction phase.

This is truly a time for celebration. Another scope of services and, more importantly, another budget is in the making. The junior engineer you put on her first project has learned quite a bit on the job without any investment in continuing education or on-the-job training programs. In fact, that junior engineer designed the only new process employed on the project. Since no one else knew how to design the new process, the junior engineer might as well have done it. In this way, the more senior engineers could design elements of the project they had designed many times before on other projects thereby avoiding any expansion of their design capabilities. In fact, many of the project specifications were identical to specifications for other earlier projects. Through the magic of word processing you have been able to charge a number of clients for the same "engineering" work.

Certainly she is now qualified to serve as a construction project engineer, although you are quite certain she has never seen a construction site. Then again, during your own six years of undergraduate and graduate engineering schools, you also were never shown a construction site. Nevertheless, twenty years ago you did quite well when assigned to your first construction project without any experience. If she makes a few mistakes, only the contractor knows about them. Anyway, who is going to believe the contractor? As is your practice, you will make sure that the client knows only you are looking out for the project's best interests. Soon your client will realize anything a contractor has to say is untrustworthy because contractors are always out for the client's money.

In no time at all, addenda are issued that you hope correct the design deficiencies that were not picked up in the cursory quality control review. The last addenda to issue related to the prebid conference during which you answered the contractor's questions off the top of your head. Whether those answers were consistent with the plans and specifications are of little concern since the plans and specifications always control regardless of what you might have misrepresented.

You did notice that certain attendees at the prebid conference represented less than desirable contractors with reputations of being claims conscious. You will have to make a mental note to carefully check their bids for any non-compliance with the bidding formalities. If one of them bid low, you will need some justification for disqualification.

Fortunately, the bids are all clean and the lowest bidder is considered one of the better contractors. You are a bit concerned, however, that the low bid was so much lower than the other bidders causing you to pause and wonder what may have been missed. Your immediate concern, however, is that your own estimate was even lower than the lowest bid and the client is going to wonder whether the project budget is adequate. You explain to your client that your estimate is indeed accurate and that the low bidder can surely perform the work within the client's project funds plus some contingency. This, of course, avoids explaining the necessity of securing additional funding beyond your estimate on which the client and funding agencies relied. At this point, you realize that change orders, which are always difficult to explain, will be more difficult on this project. Not to worry, you will be working with a "good" contractor with whom you have just had a very cooperative and friendly preconstruction meeting. You are truly encouraged and confident that this project will be one of your most successful. Little did you know that the contractor's earthmoving subcontractor was to hit groundwater 10 feet higher than what was shown on the plans.

One week into the project and you're faced with a potentially large claim. The earth-moving subcontractor has shut down and demanded extra payment for extensive dewatering. The general contractor calls for a meeting which you attend. You are told the project is being delayed and that a change order is necessary. The client wants to know how this could happen so quickly.

You must do the right thing. Go to the funding agency, explain that the groundwater problem is an unforeseen site condition that was not caused (even if not detected) by your engineering. Armed with a funding increase for the dewatering costs, the client is less likely to lose confidence in you. As for the related delays, which the funding agencies have refused to pay for, negotiate with the contractor. It is still early on in the project and you suspect that the contractor would not want to risk the good relationship you have by pressing for a delay claim. Sure enough, the contractor gets the message and backs off in anticipation of your giving him a break as the project proceeds.

It doesn't take long before the contractor wants to cash in on the favor owed. A substitute for specified equipment, surely cheaper in price, has been proposed by the contractor to the junior engineer through a shop drawing. Unlike earlier shop drawing submittals, this substitution request is going to take some thought and a simple marking of *no exceptions taken* will not be enough.

The substitute proposal relates to key project equipment and the junior engineer has the good sense to bring it to your attention. What the contractor is proposing appears reasonable even if you have never heard of the manufacturer being proposed. Unable to consult with the more senior designers who had shied away from preparing the specification at issue, you ask for an opinion from the junior engineer who performed the original design work. She tells you that the manufacturer being proposed had not provided her with any technical information during design, nor did she meet with the manufacturer's local representative. In short, the junior engineer is of little help and you are left to rely on the contractor's word. Knowing that the contractor is fully responsible for substitutions per the specifications, and feeling a bit uneasy about the earlier favor regarding the site work delay, you take no exceptions to the substitution submittal.

As the project progresses, the good relationship with the contractor is rapidly deteriorating. Your resident engineer is unable to take care of the multitude of conflicts and inconsistencies within the plans and specifications by working with the contractor's superintendent in the field. Change orders are being demanded, usually without regard to the procedural requirements set forth in the construction contract. While most of the change order demands remain pending, the contractor presses ahead just to get this project, which has turned into a losing proposition, finished as soon as possible. Then all hell breaks loose.

The substitution, the one designed by the junior engineer and substituted by the contractor, fails dramatically. The cost to correct the problem is substantial. Your client is now angry and has refused to pay the contractor. The contractor, in turn, has hired a law firm to prepare and prosecute a claim against your client. The claim is on its way and your project team is taking on a bunker mentality. The claim arrives and it includes every possible cost, both real and imagined, of the contractor, its subcontractors and suppliers. Not surprisingly, the claim includes charges for the original delay and all of the "claims" that you understood were resolved quietly in the field.

You now finally turn to the contract language. Armed with contract requirements totally ignored up to this point by you and the contractor, you attempt to nip the claim in the bud. You send the contractor a decision that reads: "The contractor's claim is hereby denied because the contractor has failed to give notice and otherwise failed to follow the procedural requirements of the contract. Furthermore, the contractor has accepted payment waiving its claim." You realize that such a determination may not be fair, but the contract gives you the right to resolve disputes between your client and the contractor and you just resolved this one. The contractor gets your "good faith" decision and walks. A lawsuit will soon follow.

The above hypothetical greatly exaggerates unethical behavior on the part of the consulting engineer. Despite the exaggeration, those who have practiced as project engineers and project managers will surely recognize how the present practice of design and construction engineering is, from time to time, subject to the type of abuse so rampant in the hypothetical.

Engineering malpractice rarely goes unnoticed, as it impacts so many project participants. Mistakes, most of which lead to disputes, can rarely be hidden or explained away. As Herbert Hoover once said,

> The great liability of the engineer compared to men of other professions is that his works are out in the open where all can see them. His acts, step by step, are in hard substance. He cannot bury his mistakes in the grave like the doctors. He cannot urge them into thin air or blame the judge like the lawyers. He cannot, like the architects, cover his failure with trees and vines. He cannot, like the politician, screen his shortcomings by blaming his opponents and hope that the people will forget. The engineer simply cannot deny that he did it. If his works do not work, he is damned forever." (Hoover 1961).

The key to successful practice then is to avoid and resolve mistakes and disputes that so often are there for all the world to see.

II.

Dispute Avoidance

Dispute avoidance is not something that begins at ground breaking. Dispute avoidance is not a technique that helps you resolve the inevitable problems that arise during construction. Dispute avoidance is not extra special contract language disclaimers strategically located in your professional services agreements and special conditions of the construction contract.

Dispute avoidance is the necessary consequence of the responsible practice of the business of consulting engineering. Dispute avoidance occurs when the consulting engineer accepts and addresses professional responsibility, not disclaims or limits potential liability. Dispute avoidance should be practiced from business and personnel development through contract preparation, quality control and performance.

1.0 Business Development

Consulting engineers, like many other professionals, sell time and only time. Although a client may view the engineer's reports, plans, specifications or other documents as products, the truth of the matter is a client is purchasing an engineer's expert attention to address a particular need of the client. Any misconception a client may have as to what an engineer provides, is fertile territory for later disputes. Unfortunately, the marketing of most engineering services promotes many misconceptions.

Business development, in one of its most basic manifestations, is illustrated in Figure 1. A marketing manager, utilizing basic tools of marketing such as brochures, presentations and firm reputation, continually seeks to bring in new business. The new business works to offset the loss of business from dissatisfied existing clients who elect to hire a competing engineering firm. The overall purpose of business development is to minimize any loss of existing client base and to achieve the desired workload by a blend of repeat client work and new client work in accordance with a carefully thought out business plan. As most marketing managers will admit, what is stated in theory rarely works in practice.

Instead, business development faces ever growing pressure to meet ambitious business plan goals in an increasingly competitive marketplace of more engineers and fewer clients. This situation is, of course, made more severe in

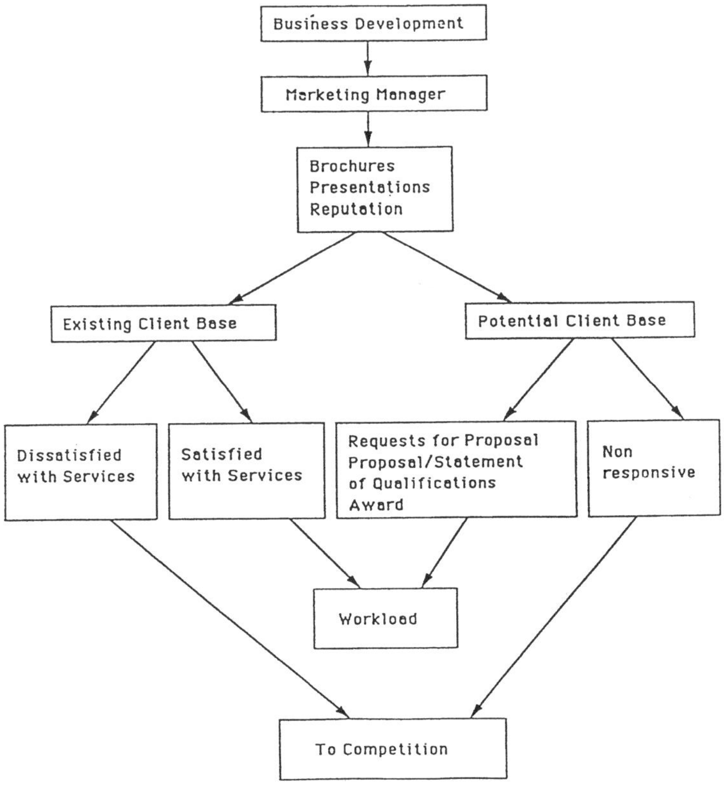

Figure 1. Business Development

tough economic times. The marketing manager is faced with aggressively marketing what the firm "sells" and how great the "product" is in order to maintain workload. In effect, business development promises the existing and potential client base more than the engineering firm is capable of delivering and, not surprisingly, clients expect what was promised.

It is important to note that the law holds professionals to a standard of practice reasonably expected of other professionals under similar circumstances. Hence, as a general rule, an engineer is not expected to perform flawless design but is expected to perform services in a manner consistent with the quality of work prepared by a reasonably prudent engineer under similar circumstances. When is the last time any marketing manager was able to sell services to a new client by boasting "we do our work just as good as the next guy"?

Instead, the marketing team promotes, by way of brochure, presentation or reputation, that the services are better, more timely and more cost-effective than the competition. The clients rely on these representations and expect a superhuman engineering effort that no engineering firm can deliver. Depending on the circumstances, the law may impose that superhuman engineering effort standard when otherwise one need only to meet a reasonably prudent engineering standard. The seeds of dispute are sown well before the project work is landed.

What can be done to avoid this potential problem?

1. Make sure the marketing manager is in touch with reality. This manager must know the capabilities and limitations of the firm's services. No engineering firm can be everything to every client.
2. The marketing manager must be familiar enough with the practice, having practiced him or herself, to know how much certain services cost and how long these services take to perform.
3. The business plan must be realistic. Overly ambitious business plans make for overly ambitious promises to clients that cannot be kept.
4. Firm brochures must not be misleading. Do not represent anything in a brochure that cannot be explained before a jury under oath.
5. Pass over projects for clients who have unrealistic expectations of the services performed and costs thereof. Let the competition inherit the difficult clients.

The system by which engineering services are typically procured deserves special note. Figure 2 depicts a fairly standard engineering services procurement system. Its overall purpose is to select the most qualified firm to meet the project needs. In theory, the client issues a Request for Proposal (RFP) inviting

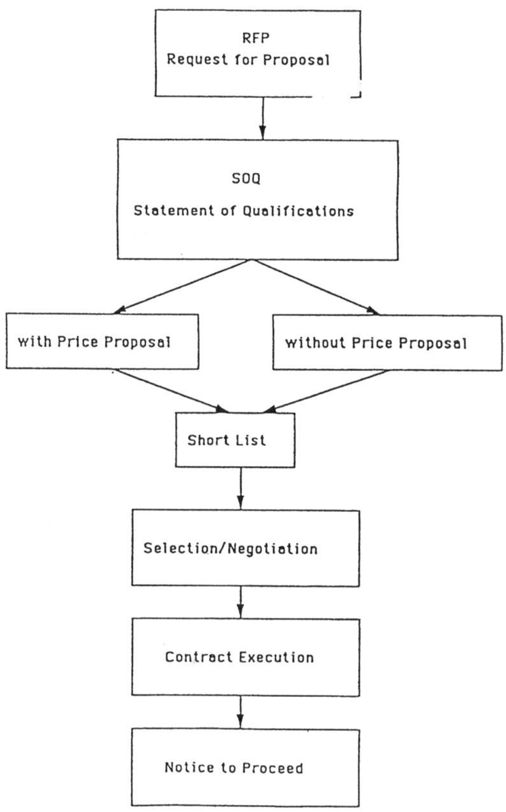

Figure 2. Engineering Services Procurement

interested engineering firms to submit a Statement of Qualifications (SOQ) with or without a Price Proposal.

All RFP's, however, are not made equal. In fact, some RFP's are the consequence of a client simply going through the motions of "competitively" evaluating engineering firms to meet a formality prior to selecting the firm that the client has already decided to select prior to issuing the RFP. Obviously, if the potential project is already in a competitor's "bag," don't waste the effort in preparing a response to the RFP.

Also, carefully evaluate the RFP for ambiguous requirements or unrealistic limitations. For instance, Figure 3 is an example of a RFP later that spells disputes. Note that the RFP calls for a Price Proposal. Prior to a 1978 decision by the U.S. Supreme Court in *National Society of Professional Engineers v. United States* (435 U.S. 679), the submission of price proposals that effectively allowed for competitive bidding was considered a violation of the Engineer's Code of Ethics. The Supreme Court found this aspect of the Code of Ethics to be an illegal restraint of trade.

It should be noted that this decision did not prevent engineers from refusing to bid professional services yet price competition between engineers has come a long way since 1978. In today's practice, most technical services procurements are heavily based on pricing, as if an engineer's services are a fungible commodity. Possibly, the impetus to bid engineering services comes from the close relationship engineering enjoys with construction, which has been traditionally bid and awarded to the lowest responsible bidder. Engineering, however, is a profession requiring as much exercise of judgment and specialized expertise as the practice of medicine or law. When is the last time the prospective client, who now insists on a lowest price proposal award for engineering services, secured surgical services of a doctor or personal injury defense services of a lawyer by lowest bid?

Hence, client education, through accurate marketing of what a professional engineer does (versus how much better and cheaper it can be done than the competition) will avoid misunderstandings and disputes. When "forced" to submit price proposals, do so as an estimate of what it costs to perform a specific type of service, not as a lump sum on an ambiguously defined project scope.

This concept is best illustrated by again turning to the RFP in Figure 3. Note that the RFP seeks not only a price proposal that will be evaluated by a selection committee but also a cap on the proposed price. This cap is imposed despite the open-ended and ambiguous nature of the scope of services. For instance,

i) "The scope of the study and report shall include, *but not be limited to*"
   – does this mean there is other undefined work required?

Request for Proposal to Perform a Regional Refuse Disposal Feasibility Study

The Town is seeking proposals from qualified engineering firms to undertake a study of regional refuse disposal and recommend a course of action.

The scope of the study and report shall include, but not be limited to:

(1) An assessment of solid waste needs in the area;
(2) Investigation of the potential of working with other nearby towns;
(3) Identification and analysis of alternative methods of regional incineration, landfilling and resource recovery;
(4) Analysis of the economic benefits of job creation at a local facility and inclusion of these economic benefits into the cost-formula analysis;
(5) Analysis of recycling and volume reduction; and
(6) Recommendations and an implementation strategy including preferred location for recommended alternative. The consultant will investigate all sources of possible funding for the design and construction phases of the project.

This study is funded by State and Federal grants and is subject to the grant requirements. Once implemented the intent of this project is to bring the town into compliance with State and Federal regulations.

The total cost of this study shall not exceed the sum of $11,000 and all work, including the Final Report, shall be completed within 12 months.

Proposals shall include:

(1) a detailed description of services to be provided;
(2) cost estimates for each of the project components;
(3) a statement of the firm's qualifications, personnel, and experience in providing similar services, and
(4) a list of client references.

Candidates will be evaluated by a Committee on the basis of (a) the extent and quality of services proposed, (b) the capability to meet the proposed schedule, (c) the firm's qualifications and experience, and (d) proposed costs. The Town reserves the right to reject any or all proposals.

Proposals shall be submitted to the Board of Selectmen, Town Office.

Figure 3.

ii) "an assessment of solid waste needs in the area"
   - how detailed an assessment?
   - how large an area?

iii) "Identification and analysis of alternative methods"
   - are incineration, landfilling and resource recovery the only methods to be evaluated?

iv) "cost-formula analysis"
   - what is a cost-formula analysis?

v) "Preferred location"
   - Is there a specific number of sites under consideration or must other sites be located?

vi) "all sources of funding"
   - must every possible private and public funding source be checked?

vii) "State and Federal grant requirements...and regulations"
   - Are these available?
   - What if they change?

viii) "$11,000"
   - is this all the funds appropriated?
   - if the project exceeds this price is the Town obligated to appropriate more funds?

With these and many more questions, you could call the Town Offices, likely reaching secretarial staff who are not authorized to divulge more details over the phone or in writing. What can be done?

The proper and dispute-avoiding response to such a RFP is to offer a scope of services and related price schedule as *detailed as possible.* Define exactly what will be performed for the price offered. Do not be all inclusive, be exclusive. Essentially, make a detailed counteroffer which, if accepted, becomes the essence of the contract with the client. To accept an ill-defined open-ended scope of services of a RFP is to guarantee client disputes and an inadequate project budget.

With respect to the Statement of Qualification, identify those technical personnel that will be working on the project. Common client disputes arise when the best engineers are proposed, but "second stringers" show up when the project commences. Similarly, when short-listed for a presentation, attend the presentation with representatives of the technical team who will be working on the project. In this way, the client comes away with a *realistic expectation of who is going to do what for how much and when.* A key ingredient of any dispute is when one party's expectation differs from another's intention. Make sure what you intend to provide is what the client expects to be provided.

Finally, with respect to selection/negotiation there is little to add if a price proposal has been submitted for a detailed scope; the only contractual language that remains is administrative and best left for legal counsel. If a price proposal has not been provided, negotiations can clarify for the client the project scope and establish a reasonable fee. If the prospective client insists on defining the scope and an unrealistic fee, turn down the work. This is always easier said than done but in certain instances, the best dispute avoidance technique is to refuse to provide the services.

1.1 The Hypothetical Revisited

The hypothetical seasoned engineer had sown the seeds of dispute before the project was secured. The seasoned engineer marketed his firm's capabilities having been out of touch with project work himself and quite likely out of touch with his firm's capabilities. It is always difficult to sell that which you do not do or know.

The seasoned engineer also sold the client on price, an easy sell, versus expertise, a sale for which the engineer had to make the effort to educate the client as to the practice of engineering. Ultimately, the seasoned engineer came away with an unrealistically budgeted project and the client came away with an unrealistic expectation of what the engineer can and would do. Before the project began, a dispute was simply unavoidable.

2.0 Personnel Development

Consulting engineers sell time and the quality of that time is directly dependent on the talent of a firm's personnel. In other words, the most important investment a firm can make is in its personnel. For some reason, the investment many consulting engineers make in their own talent is overshadowed by the more tangible investments made in marketing, computer assisted design capabilities or office facilities. Yet there is no dispute avoidance technique as effective as high quality planning, design and construction administration services provided by a talented engineering staff.

Some firms have no formal personnel development program. They rely solely on the initiative of its staff to make the satisfactory degree of effort on each project assignment, keep up-to-date on advancing technology and qualify for the appropriate credentials. Few firms have an actively enforced formalized program that supports advanced training, carefully oversees the performance of each assignment and records each engineers' efforts with an eye towards advancement within the organization. Most firms have a formalized program, more often haphazardly followed, if followed, at all. It is to this majority that lawyers can most easily find fault when pursuing a dispute.

There is little more damaging to the defense of a consulting firm than the systematic identification of internal personnel policies that were either ignored or breached. If the policies require registered professional engineers in certain disciplines to do certain work, do not assign someone without the credentials. If the policies require yearly continuing education, make sure the staff takes this continuing education. If the policies require periodic reviews of work quality, perform the reviews and correct any deficiencies found. In its most simple terms, if management decides to define a personnel policy, then management should make sure that policy is fully implemented. Otherwise, litigation and creative lawyering can result in both damaging and embarrassing consequences. For instance, a cross-examination may read as follows:

Q: You are the project manager on the project, is that right?
A: Yes.
Q: You consider yourself to be fully competent?
A: That's right.
Q: And you consider your firm to be competent and professional?
A: Absolutely.
Q: And your firm's practices are in keeping with competent engineering, isn't that true?
A: Yes.
Q: The technology you deal with is always advancing, isn't it?
A: Certainly.
Q: In fact, you get paid for your expertise on this advancing technology, don't you?
A: Yes, indeed.
Q: You would consider it important that you and your staff keep up to date on advancing technology?
A: I do.
Q: And your firm has a policy regarding training for this very purpose?
A: I believe so.
Q: Is this that policy? (handing the witness the manual of practice)
A: That's what it appears to be.
Q: Don't you know?
A: Yes, that's our manual.
Q: Turning to the manual at the chapter identified as "qualifications," do you see where it says that the project manager must have 10 years experience to be assigned to a project of this kind?
A: Yes.
Q: You don't have 10 years experience, do you?
A: No.
Q: It also says, all project managers must have a Master of Science and take 20 hours of continuing education per year of practice. You don't have a Masters of Science, do you?
A: No.
Q: When is the last time you took a continuing education course?
A: It has been awhile.

Q: It also says, project managers must serve as engineers on similar projects before being promoted to the project manager position, do you see that?
A: I can read.
Q: What similar projects were you assigned to as a project engineer?
A: None, this is my first project of this kind.
Q: You'd have to agree with me then, that you don't have the experience or education required by this policy to be a project manager.
A: The policy is just a guideline and not always followed.
Q: And, in your case it hasn't been followed, true?
A: Apparently so.

2.1 The Hypothetical Revisited

Our seasoned engineer believed the best training is on-the-job training. Although there is no substitution for real project experience, the client should not pay the tuition. Here, the junior engineer was given considerable project responsibility without any training and little supervision. The seasoned engineer knew from his own experience that junior engineers, fresh out of engineering school, are green with incompetence.

Instead of assigning the junior engineer to work with a more experienced professional on a less complex aspect of the project, the junior engineer took on a difficult assignment and the more experienced engineers retreated to the comforts of past design experiences. It is those very past design experiences that could have been drawn on had someone other than the most junior engineer been assigned the more complex portion of the project.

The sink or swim mentality of a junior engineer's introduction to the profession can and does lead to hard lessons for all.

3.0 Quality Control

Quality control is the one area where breach of internal quality control policies can be *more* damaging than breach of internal personnel policies. With an effective quality control program, disputes can be avoided because mistakes are caught before any damage is done. Again, because of time or budgetary restrictions, many projects do not receive the quality control checks, interdisciplinary coordination checks, or "buildability" evaluations necessary to catch the inevitable design errors. As previously mentioned, an engineer is *not* expected to produce a perfect design, but a disgruntled client will always argue that the imperfection that caused the problem would have been caught had quality control procedures been followed. For example:

Q: You have a quality control program relating to plans and specifications, don't you?
A: The company does.
Q: Is this that program? (handing the witness the quality control program)
A: Yes it is.
Q: The purpose of the program is to catch mistakes, isn't it?
A: That's true.
Q: If followed, the program would catch problems with the plans and specifications, right?
A: Supposed to.
Q: If not followed, mistakes in the plans and specifications would not be caught, true?
A: Maybe.
Q: The quality control program wasn't followed in this project, was it?
A: Not entirely, we had a budget problem.
Q: You'd agree with me then to the extent it wasn't followed, mistakes in the plans and specifications were not caught?
A: True, if there were mistakes in the first place.
Q: Is your work perfect?
A: Nobody's perfect.
Q: Were the plans and specifications perfect?
A: No plans and specifications are perfect.
Q: All plans and specifications have mistakes, right?
A: Yes.
Q: If you had employed the quality control program, you would have caught mistakes, isn't that right?
A: Maybe.
Q: We'll never know in this case, will we?
A: No.

Internal quality control policies can effectively define the standard to which you must practice. In legal jargon, the policies may be evidence of the standard of care and a breach of a policy may be evidence of a breach of the standard of care, i.e., negligence. If there is justification for not following a quality control procedure, document that justification. If called to explain why certain procedures were not followed, there will be, at least, a documented reason.

Quality control, in its broadest sense, can encompass programs that improve the performance of services as well as the checking of results. Value engineering, peer review and total quality management are examples of quality control programs that look at how a design is delivered instead of checking the design product.

3.1 Value Engineering

Value engineering looks at alternative ways to meet a project's requirements in a more efficient and cost-effective way.

A typical value engineering program would provide expert professionals the opportunity to "brainstorm" during structured sessions. The value engineering sessions would then identify a series of design alternatives that may have escaped the project team's attention. These alternatives are then ranked against criteria ranging from the more objective, such as construction cost or operation and maintenance costs to the more subjective, such as aesthetics or the risk of innovation. The alternatives are then ranked overall and, where appropriate, implemented into the final project design.

Value engineering's greatest asset is its opportunity to give a fresh look at a project thereby improving the quality of the end product.

3.2 Peer Review

Unlike value engineering, which focuses on improving an engineered design, peer review evaluates the management and operation of the engineering firm. Business objectives, operating policies, human resources, financial management, and business development are each assessed in a peer review program. A critical assessment is made on how management is structured and operated, as well as how project teams perform and quality is controlled.

Typically, a peer review program involves management from other non-competing engineering firms of near equal size and practice participating on a peer review team that assesses the firm's practices. The peer review team then provides recommendations as to improving practice.

The peer review critique is obviously sensitive. Yet, despite its confidential nature, there is no assurance that the results of peer review (or for that matter, the results of value engineering) would not be disclosed in an adversarial proceeding. In response to this possibility, many peer review programs encourage the destruction of all records after an assessment is complete. This does not guarantee confidentiality since the peer review team itself is subject to examination of all "discoverable" evidence. The standard of what is discoverable is quite broad and would surely include peer review assessments under the right circumstances.

Hence, like quality control, if value engineering or peer review practices are employed, be fully prepared to implement the resulting recommendations or, at the very least, justify why such recommendations were not followed.

3.3 Total Quality Management (TQM)

Total quality management or TQM is the most recent quality control tool to be applied to the consulting engineer's practice. TQM is more of a management philosophy than a quality control technique in that TQM focuses on improving performance at all levels of the engineering firm.

TQM originated from manufacturers seeking ways to improve production processing and advancing production technology. As applied to engineering services, TQM emphasizes the process of producing quality design and services while focusing on the importance of the employees' participation in this process. Hence, TQM looks at improving the system that provides the engineering services versus checking the engineering services themselves.

A TQM program, as applied to professional service organizations, focuses on:

1. **The supplier/customer.** TQM fully integrates the concept that customers are not only the clients, but also the employees in that they provide services to each other. By emphasizing the importance of meeting the customers' needs at all employee levels, the overall end product's quality should improve.
2. **Process/Results.** Most quality control programs focus on checking the quality of the resulting service or product. TQM looks at the processes by which the service is provided or product produced. If one improves the process, the quality problems are prevented - a far more cost-effective approach than checking for errors.
3. **Feedback.** The employees providing the services have enormous expertise that, if tapped, can greatly improve the process of providing services. TQM encourages employees to share this expertise and to correct problems at all various levels of the organization.

Applying TQM to consulting engineer management is a drastic change to the conventional results-oriented style of management common to most firms. Yet, TQM is a management philosophy that can greatly improve the quality of the engineering services to the client in a cost-competitive and efficient manner.

Keep in mind that quality control, value engineering, peer review and TQM are important techniques in improving quality and avoiding disputes. Failure to follow through on these techniques, however, can be and usually is more problematic than if the techniques had never been implemented in the first place.

3.4 The Hypothetical Revisited

The hypothetical takes the most ill-advised approach to quality control—that is quality control is expendable in the face of budgetary problems. By providing cursory review of a design, our seasoned engineer is asking for later trouble.

Addenda is not the way to correct design errors. Construction is not the time to coordinate drawings. No design is ever perfect and addenda will be necessary to pick up oversights just as change orders will be necessary to correct problems discovered in the field. Neither addenda nor field corrections, however, should be expected to correct that which could and should have been corrected during design.

In part, because of the manner in which the project was marketed and the budget established, the seasoned engineer has cut back on quality and assured design defects.

4.0 Professional Services Agreement

The days of the handshake contract are long gone. Unfortunately, some traditions die hard and many consulting firms continue to allow informal contract preparation (i.e. letter agreements, letters of understanding, etc.). Possibly, this is for fear of questioning the integrity of long-term clients. More likely, it is just easier than dealing with complicated forms, or worse, lawyers.

Some engineers not only continue with informal agreements for professional services, but also fail to limit the authority of its employees from entering into or changing preexisting agreements. Hence, a project engineer or resident engineer who has already developed a working relationship with a client may perform services requested by the client without the formality of having the original scope of services revised. Although this practice may be against a firm's policy, it is also convenient and contractually dangerous. Furthermore, it could leave the engineering firm liable for serious risks not contemplated during the initial agreement and well beyond the risks normally associated with the work. Management must limit the authority to enter or modify contracts to a select few, enforce that policy throughout the firm, and make sure clients understand this limitation.

Regardless of the complexity of the form of agreement, the contract for professional services is the single most important document for avoiding disputes if it includes the following necessary and basic elements;

1. a clear and precise description of the services to be performed,
2. the time frame within which the services are to be performed, and
3. the cost of the services including the method of payment and the timing of payments.

There are a number of standard contract forms prepared by professional societies, including the Engineers Joint Contract Documents Committee (EJCDC) documents as well as the American Institute of Architects' (AIA) contract documents. Although these contract forms are excellent contract models and quite comprehensive in scope, they are no substitution for the careful formulation of the above three items. Also, standard forms of agreements, whether created in-house or taken from the EJCDC or AIA, may not be appropriate for unique project requirements. It is not unusual to find standardized forms applied to projects for which they are simply not appropriate. Standard forms should be customized for each project after consultation with legal counsel.

As noted in the warning printed on the forms themselves "**This document has important legal consequences, consultation with an attorney is encouraged with respect to its completion or modification.**"

In addition to the basic and necessary elements of the professional services contract that act to avoid disputes, there are a few other provisions that are helpful:

1. A provision that defines the additional services for which the engineer is entitled for additional compensation. This, again, goes to clarify the scope of services.
2. A merger clause that effectively states that all prior written or oral understandings are superseded by the agreement. This type of clause addresses the "over ambitious" marketing effort.
3. An ownership of documents clause that retains the engineers' ownership of its work product to avoid a client's later reuse (and possibly misuse) of work product.

Provisions that disclaim or limit an engineer's liability are generally not accepted by clients and are difficult to enforce in court. More importantly, such clauses are only effective against the client and not against third parties, such as contractors, who are more likely to become involved in disputes with an engineer. If the project risks call for such provisions and the client agrees to them, by all means include such provisions in the services agreement. Do not, however, rely on such provisions offering unlimited protection from disputes.

As a final note, do not agree to provisions that "guarantee" your work or set the quality of your work above that which you are required to meet. More importantly, do not agree to be held responsible for your client's or any other party's failing. For instance;

> The engineer shall guarantee the plans and specifications to be free of errors and omissions and, in the event an error or omission is discovered, shall correct the same at no additional costs. or
>
> The engineer shall perform professional services in accordance with the highest professional standards. or
>
> The engineer shall indemnify and hold harmless the owner from all liability that arises from the acts or omissions of the engineer and owner during the course of the project.

To guarantee perfection of work product or the performance of services is to set the stage for major disputes. To indemnify another is to effectively get into the insurance business for which engineers are neither paid nor qualified. Furthermore, the cost to resolve disputes that arise from such provisions may rest entirely in-house, since errors and omissions insurance policies usually exclude coverage of this kind without riders to the contrary.

4.1 The Hypothetical Revisited

Our seasoned engineer has taken comfort in a letter agreement that is vaguely written. With a poorly defined scope, the seasoned engineer argued that the scope has been exceeded and the fee should be increased. The uninformed (and now surely suspicious) hypothetical client bought this argument and increased the engineer's fee, but at what cost.

To avoid disputes, the client must fully understand what services are to be provided from project initiation through performance. A fully detailed engineering contract and scope of services goes a long way in reaching this understanding. Our seasoned engineer's vague agreement may become fodder for attorneys who, because of this vagueness, can argue what the contract really means and who was responsible for what. When, or if, these arguments came to pass, our seasoned engineer would likely be on the losing end.

There are certain basic rules of contract interpretation. One rule is that an ambiguous provision in a contract is interpreted against the author. Another rule is that ambiguous contract terms may be interpreted in the context of pre-existing representations (such as marketing efforts). Here, given the promises of highest quality and extensive services with a vague scope of services, the engineer's letter agreement would likely be interpreted to include, instead of exclude, services. Such a dispute would have been avoided had our seasoned engineer and client negotiated a clear, comprehensive, and fair professional services agreement.

5.0 Performance of Services

The key services provided by consulting engineers usually involve design and construction administration services including:

- Preparation of Administrative Provisions of Contract Documents
- Preparation of Plans and Specifications
- Construction Cost Estimates
- Bid Evaluation/Award
- Preconstruction Meeting

- Substitutions
- Shop Drawing Review
- Scheduling of Construction Activities
- Construction Inspection
- Change Order/Claim Evaluation
- Payment Certification
- Safety Considerations

Each of these services can and should be performed with due regard to avoiding disputes, yet many of these services are performed in a manner that promotes disputes.

5.1 Preparation of Administrative Provisions of Contract Documents

General Conditions, Supplementary Conditions, Special Conditions, Agreement and other related administrative provisions are prepared by engineers along with the technical specifications of most construction contracts. These administrative provisions, however, are completely different in scope and impact on the rights and obligations between the parties than, for example, a piping specification. Engineers have the expertise to prepare technical specifications, but rarely have the expertise to fully deal with the **legal consequences** of administrative provisions; yet, by tradition or default, the preparation of administrative provisions has come to rest with the engineer.

Typically, the engineer will rely on an EJCDC or AIA published standard or, possibly, an in-house standard that has proven effective on earlier projects. Many public client-owners also have standard provisions.

These standards, like those published for professional services agreements, are excellent and comprehensive resources. The use of these standards in the preparation of contract documents between parties other than the engineer (i.e., owner/contractor), however, has entirely different consequences than using standards for a professional service agreement. For instance, engineers are not trained in contract law, indemnification, alternative dispute resolution, and the host of other terms and provisions that define the rights and obligations between the parties. Since ignorance is never a defense, an engineer who enters into a professional services agreement defining obligations neither contemplated nor understood will nonetheless be bound by these obligations. Hence, the consequence is to the engineer. When this same engineer prepares contract documents for an owner and general contractor, it is the owner and contractor who are bound by provisions imposed by the engineer and outside the engineer's expertise. Not surprisingly, this is a common cause of later disputes with the engineer.

This distinction between preparing one's own versus another's contract is important. One is always free to negotiate the terms of one's own contract. Providing advice, however, on the terms of a contract to be entered into by another, is the practice of law, not engineering. Yet there is no question that most engineers take on the practice of law by preparing administrative contract document provisions for their clients with little hesitation and usually less thought than what goes into technical specification preparation.

It is interesting to note that the professional societies representing engineers, architects and lawyers have recognized this problem. The AIA, in its Handbook of Professional Practice states:

> The standard AIA documents are prepared under close scrutiny of AIA's legal counsel, who meets regularly with the AIA Documents Committee and staff. Still, the architect should not advise the owner or anyone else on legal matters even though the advice may involve no more than the use or modification of the standard AIA contract documents. A layperson advising others on contract matters is engaging in unauthorized practice of law.

A number of things can be done to avoid disputes in this area.

1. If the client has his or her own contract administrative provisions, use them in their entirety without modification.
2. Educate the client. Explain that engineering services are technical services and that the preparation of non-technical provisions of the construction contract documents are beyond this expertise.
3. If the client expects a complete contract, including administrative provisions, use a standard form and ask the client, in writing, to have legal counsel review and complete the provisions.

One of the key legal consequences of the administrative provisions of the construction contract is the allocation of risk between the owner and contractor. Engineers, having been empowered to draft or revise these provisions, have the understandable tendency to include provisions that shift most, if not all, of the risk to the contractor. This appears to be a noble effort on behalf of the client-owner. However, on a purely practical basis, the unfair allocation of risks on a construction project promotes, instead of discourages, later disputes.

Examples of the more common contractual risk-shifting clauses found in construction contracts and generally accepted as standard practice are illustrated on Figure 4. These clauses transfer risks that contractors can control. Examples of less common and inherently unfair clauses imposed on contractors, who are in no position to negotiate more equitable provisions under common bidding practices, are depicted on Figure 5. These later clauses transfer risks that simply cannot be controlled or anticipated by the contractor. In some instances, the risk is

Examples of Common Risk Transfer Devices

1. Allocation of Safety Responsibility to the Contractor
   10.1 Safety Precautions and Programs
      10.1.1 The Contractor shall be responsible for initiating, maintaining, and supervising all safety precautions and programs in connection with the performance of the Contract.
   10.2 Safety of Persons and Property
      10.2.1 The Contractor shall take reasonable precautions for safety of, and shall provide reasonable protection to prevent damage, injury or loss to:

      (AIA Document A201–1987, § 10.1 and 10.2)

The Contractor is in the best position of promoting site safety and should be contractually responsible for the same.

2. Allocation of Supervision Responsibility to the Contractor
   3.3 Supervision and Construction Procedures
      3.3.1 The Contractor shall supervise and direct the Work, using the Contractor's best skill and attention. The Contractor shall be solely responsible for and have control over construction means, methods, techniques, sequences and procedures and for coordinating all portions of the Work under the Contract.

      (AIA Document A201–1987, § 3.3.1)

As with site safety, the Contractor is in the best position to supervise the project and should clearly be made contractually responsible for the supervision.

3. Clauses Which Negate Any Duty of the A/E to Supervise Construction Or To Conduct Safety Programs
      2.6.6 The Architect shall not have control over or charge of and shall not be responsible for *construction means, methods, techniques, sequences or procedures,* or for *safety precautions and programs* in connection with the Work, since these are solely the Contractor's responsibility under the Contract for Construction. . . .

      (AIA Document B141–1987, § 2.6.6)

This type of clause is more of a risk limiting device in favor of the architect but also acts to clearly define means and methods responsibility as the Contractor's.

Figure 4.

shifted to the contractor with control vested in the engineer or owner, a very severe proposition and sure to be disputed if enforced.

Despite clear contract language of risk-shifting clauses, courts hesitate to impose liability on otherwise innocent parties. Hence, when challenged, many of these clauses are found void by the Courts and offer little protection against claims.

One type of common risk-shifting or, more accurately, risk-setting clause found in many construction contracts is the liquidated damages clause. This clause sets the daily damages to be imposed because of late completion, thereby avoiding a dispute as to how much the owner is actually damaged. For example:

Example of Severe Risk Transfer Devices

1. No Fault Forfeiture Clauses
"The Contractor shall not permit hazardous waste to be incorporated into or generated by the Project"
This type of clause obligates the Contractor to insure something by contract that would not otherwise be imposed by law and may be completely outside the Contractor's control.

2. Perpetuating the Contractor's Responsibility
"Contractor shall be responsible for the consequences of the revocation of the Engineer's approval because of the failure to identify portions of non-conforming shop drawings."
Clearly, the Contractor cannot rely on the Engineer's approval under these circumstances as the risk of revocation of acceptance remains.

3. Limiting the Contractor's Damages
"No payment of any kind shall be made for damages because of delay from any cause in the progress of the work."
Hence, even if the contractor is not responsible, the right to recover delay damages has been lost by Contract.

4. Shifting the Engineer's Responsibility
"The Contractor is responsible for all design to accommodate the substitution of any product not contemplated by the Engineer."
These clauses essentially abdicate the Engineer's design responsibility to an unqualified Contractor.

5. Shifting the Owner's Responsibility
"The Contractor shall be responsible for the continued and uninterrupted operation of the owner's facilities during construction."
The Contractor is probably unqualified to operate the owner's facilities, yet is made responsible for their continued and uninterrupted operation.

Figure 5.

The Contractor shall pay to the owner the sum of $500 per day for each day that the Contractor fails to complete the project within the time set forth in the contract.

Liquidated damages clauses can be very effective in avoiding disputes as to the consequences of late completion of the project so long as the liquidated damage amount is properly established. Unfortunately, many engineers arbitrarily set liquidated damages as part of the preparation of the contract administrative provisions and view this provision as a penalty clause for late completion.

As a general principal, courts will not enforce a penalty in a contract even if the parties agree to the penalty term. Courts will enforce a liquidated damages provision so long as the amount set, at the time the contract was formed, reasonably relates to the amount of damages anticipated. In other words, as the engineer, you should estimate what it would cost in the event of late completion, document that estimate, then set the liquidated damages amount accordingly. In many instances that cost may be engineering costs during construction only. Do not simply pick whatever number you believe will give the contractor enough incentive to complete the project; a reasonable estimate of actual damages is enough incentive and anything above that amount is likely unenforceable.

In many construction contracts, the responsibility-shifting clauses are coupled with authority-enhancing clauses, for instance

> The Engineer shall be the final interpreter of the plans and specifications whose decision is final and binding on the Contractor.

This typical clause creates one of the most troublesome conflicts of interest in engineering practice. By this clause, the engineer has become judge and jury as to his or her own work product. When called upon to make a decision between differing contract language interpretations there is a natural tendency to favor that interpretation which avoids cost to the client-owner and any suggestion that the plans and specifications were somehow faulty. Contractors have come to expect an engineer's decision to be adverse even when an engineer is obligated to make impartial decisions.

Given the power to allocate risks by enhancing the engineer's or owner's authority while shifting as much responsibility to the contractor as possible, few engineers exercise restraint. Construction contracts that are unfairly one-sided in this regard **do not** offer protection from disputes. In more cases than not, contractors cannot adjust bids in any realistic fashion to reflect unbalanced risk allocation and remain competitive. Furthermore, once burdened by responsibility not based on fault, most contractors will vigorously protest and commonly litigate disputes. Keep in mind, there are very few risk allocation clauses that cannot be challenged by imaginative legal counsel.

The engineer, in preparing the administrative provisions of the construction contract, should temper the inclination to shift onto the contractor responsibilities that cannot realistically be priced, are not fault related, and will likely result in a dispute if imposed during the project performance.

5.2 Preparation of Plans and Specifications

Technical specifications and plans are the heart of any construction contract. The preparation of these specifications, along with coordinated plans, are the single most important documents generated by the engineer. These documents represent the culmination of the engineer's planning and design efforts.

Most engineering firms maintain a library of specifications representing various standard specifications used on all projects as well as custom specifications prepared on past projects. Most engineers have adopted the Construction

Specifications Institute (CSI) format for specifications making specification section retrieval much easier than the more cumbersome practice of custom formatting all project specifications.

In the normal course, applicable standard and custom specifications from past projects are reorganized into a working specification for the project of interest. This working specification is then revised to accommodate the new project requirements. Custom sections are then added to cover equipment or construction unique to the project. Contemporaneously, plans are prepared which, like the specifications, include certain standard details. The majority of the plans, however, illustrate the unique requirements of the construction project.

Not surprisingly, the best dispute avoidance technique regarding plans and specifications is quality. The higher the quality of the combined documents, the fewer disputes. Although quality control with respect to plans and specifications has already been addressed, a few additional observations are in order.

1. **Avoid Mistake Repetition**. Establish a post project reporting system that identifies specification weaknesses and allows for corrections to be implemented in the specification library. The project construction period will inevitably identify specification problems. Report these problems to a central clearinghouse that implements changes in the specification library. Do not let a specification error repeat itself project after project.
2. **Draft the Specifications in Clear and Precise Language**. An example will best illustrate this point. Figure 6 is an electrical specification that purportedly defines electrical work. This specification raises more questions than it answers. For instance,
   i) What is the entire scope of work under 1.02A? More particularly, what is work that "reasonably may be implied as being incidental to this work"?

      Who makes the determination as to what is implied or reasonable or incidental?
   ii) Who are "all authorities" under 1.02B and what are all changes required to accomplish the intent? Whose intent? the engineers? the authorities? the owners? the contractors?
   iii) What happens after the request to relocate is made under 1.03A? Is the contractor responsible to relocate at his cost? How can the contractor not disturb an active service that needs relocation?
   iv) How can the contractor possibly examine "all existing conditions" under 1.04A? Does this conflict with general or supplemental conditions?

   Figure 7 illustrates certain revisions to the specification which adds clarity and delete subjective language. The specifications and drawings should

Example of an Ambiguous Specification

PART 1 GENERAL

1.01 WORK INCLUDED

A. Provide and install electrical equipment.

1.02 DESCRIPTION

A. Furnish all labor, materials, services, supplies, tools, equipment, apparatus, transportation, facilities and incidentals required and perform all operations necessary to accomplish the complete installation of the electrical power, lighting, control and communication systems as shown on the Contract Drawings, specified herein or as reasonably may be implied as being incidental to this work.

B. The performance of the work under Division 16 shall be in accordance with the regulations and rulings of all authorities having jurisdiction over the work. Any changes required to accomplish the intent of these Specifications shall be the CONTRACTOR's responsibility as to accomplishment and any extra cost for performing work.

1.03 EXISTING WORK

A. Active: If existing active services are encountered that require relocation, make request in writing for determination. Do not prevent or disturb active services that are to remain.

1.04 SITE VISITATION

A. The Contractor shall visit the site and shall examine all existing conditions which may affect his work under this contract. No claims for extra compensation or extension will be allowed because of his lack of compliance herewith.

Figure 6.

be objective criteria upon which a contractor's performance can be measured. There should be no question as to *what* is to be accomplished.

Many engineers write specifications to provide an out and to catch details missed during design. For example, Figure 6 includes language that allows the engineer to require the contractor to perform work that "reasonably may be implied as incidental" and is in compliance with "all authorities having jurisdiction." In practice, this may translate into "the contractor is required to perform that which the engineer failed to detail in accordance with those codes to which the engineer failed to refer." Such a standard of performance is, if anything, dispute prone.

Hence, the key to specification draftsmanship is clarity and precision. Do not be overly inclusive of undefined requirements. Do not refer to unidentified and possibly unrelated authorities or standards. Clearly define what work is included and what is excluded. Do not require the impracticable or impossible. Be internally consistent within the specifications and between the specifications and plans. Do not bury administrative

Example of a More Precise Specification

PART 1 GENERAL

1.01 WORK INCLUDED

A. Provide and install electrical equipment.

1.02 DESCRIPTION

A. Furnish all labor, materials, services, supplies, tools, equipment, apparatus, transportation, facilities and incidentals required and perform all operations necessary to accomplish the complete installation of the electrical power, lighting, control and communication systems as shown on the Contract Drawings *and* specified herein

B. The performance of the work under Division 16 shall be in accordance with the regulations and rulings of *the local building authority and all codes adopted thereby.*

1.03 EXISTING WORK

A. Active: If existing active services are encountered that require relocation, *notify the Engineer and proceed to relocate as directed.* Do not prevent or disturb active services that are to remain. *All such service relocation to be paid as time and material charges pursuant to the general conditions.*

1.04 SITE VISITATION

A. The Contractor shall visit the site and shall examine existing conditions which may affect his work under this contract.

Figure 7.

type provisions in the technical specifications. Do not use terms interchangeably. For instance, "sandy silt" and "silty sand" have completely different meanings and should be specifically defined and not used interchangably. Avoid vague terms such as "reasonable" or "workmanlike." Avoid indefinite instructions such as "at the direction or discretion of the Engineer." All of these suggestions go to the clarity and precision of the specifications.

As previously noted, one of the basic doctrines of contract law is that ambiguous contract language is interpreted against the drafter who in this case, is the client-owner.

3. **Distinguish Between Trades.** Segregation of trade work within the major divisions of the specifications and within the plans is simply good practice. Do not incorporate by reference or overlap trades if not necessary. In this way, there will be clearer divisions of responsibility between the various subcontracted trades and hopefully fewer conflicts and disputes as to responsibility between the trades and the general contractor.
4. **Avoid Blending Performance With Proprietary Specifications.** Pure proprietary equipment specifications typically name a certain manufacturer

and model that meets the specifications thereby making it absolutely clear to the contractor what equipment is to be provided. These specifications give the engineer prior and certain knowledge of what equipment is to be furnished and the design requirements of that equipment. Unfortunately, a named manufacturer and model does little to promote supplier price competition.

Pure performance specifications identify the performance aspects of the equipment leaving it to the contractor to select manufacturer and model. Performance specifications promote competition between suppliers but leave the engineer uncertain as to the specific equipment to be provided and how to accommodate this equipment in the overall design. Performance specifications, however, shift the burden of selecting the equipment and design to the contractor.

Some specifications blend performance and proprietary specifications. A common practice is to name a number of manufacturers and models that meet a certain specification. Some specifications allow for an "or equal" as a manufacturer. So long as the overall design is prepared to accommodate the most restrictive requirements of the named (and "or equal") manufacturer, this technique introduces an element of price competition and retains some control over suppliers.

Blending performance and proprietary specifications requires significantly more work on the part of the engineer since numerous manufacturers must be investigated. Plans and specifications that do not accommodate the worst conditions of all manufacturers and instead are prepared based on the manufacturer preferred by the engineer result in the inevitable dispute with the contractor who proposes a different manufacturer. Either the proposed manufacturer is then rejected or a significant redesign effort is required to accommodate the proposed manufacturer. If the specification must blend proprietary and performance requirements, make sure the overall design can accommodate whatever the contractor proposes. There is no avoiding extensive manufacturer investigation under these circumstances.

5.3 Construction Cost Estimates

Construction cost estimating involves the detailed determination of quantities of labor and materials applied against estimates of prevailing costs marked up by contingencies in an effort to fall somewhere other than lower than the low bid. Unlike a contractor who normally has full access to market sensitive equipment and material and labor costs, the engineer must rely on quotations from potential suppliers and standard cost data, neither of which is particularly reliable. With this limited data base, the engineer applies a contingency of having little statistical significance to reach a rough project cost. Clearly, construction cost estimating by consulting engineers is as much an art as science.

Project financing, however, is entirely dependent on the accuracy of the engineer's estimate in predicting the lowest bid. Clients come to rely on these estimates for planning purposes, grant funding applications, public money appropriations, and fiscal planning. These estimates, however, are not accurate for a variety of reasons including:

1. **The engineer cannot anticipate the bidding environment at bid opening.** The number and "aggressiveness" of bidders is highly variable, time sensitive, and can greatly impact pricing.
2. **The engineer cannot anticipate how the project is perceived by the bidders.** A high risk project will result in higher contingencies added to bids. Similarly, a project that disproportionately shifts risks from the owner to the contractor will promote added and unpredictable bid contingencies.
3. **The engineer cannot anticipate a contractor "buying" the project at or below cost.** Possibly the contractor simply needs the work or wants to break into the market and is willing to take the project at or less than cost.
4. **The engineer cannot anticipate rapidly changing wage scales or the effect proprietary specifications may have on specified material and equipment costs.**

The client must be made to understand that there is no crystal ball. No matter how exact the engineer may prepare quantity take-offs, the estimate is no more precise than the bidding process itself. If numerous competing contractors, in many instances putting the financial stability of their very businesses on the line, result in wide spreads of bids, how is the engineer to be any more accurate?

The limitations of an engineer's estimate should be in *writing*. The professional services agreement should include language such as:

> The Engineer has no control over the cost of labor, materials, equipment or services; or the Contractors' methods of determining these costs; or competitive bidding or market conditions. Opinions of construction cost are made on the basis of judgment. There is no representation that actual bids will not significantly vary from opinions of cost.

Contract language is not enough. The engineer should not give the client any reason to rely on the estimate without acknowledging the estimate's limitations. As with so many things in the litigious world of construction, put it in writing. For instance:

Dear Client:

Enclosed please find our estimate of construction cost on the project. This estimate has been prepared in keeping with ordinary engineering practice.

The estimate is not a guarantee of project costs and should not be relied on as such. We simply cannot predict how the contractors will prepare their bids at bid opening. The estimate, however, is as accurate as we can reasonably make it at this time.

We look forward to bid opening and working with you through project construction.

Very truly yours,

Engineer

The client will surely ignore the warning and if the estimate proves inaccurate, suggest that the estimate was not properly prepared. Reference to the professional services agreement language, as well as any confirming correspondence, will respond to, if not resolve, this dispute.

5.4 Bid Evaluation/Award

A competent contractor bidding on a quality set of plans and specifications guarantees a minimal number of disputes. Poor plans and specifications with a competent contractor or quality plans and specifications with an incompetent contractor spells trouble. Poor plans and specifications with an incompetent contractor spells disaster.

Bid disputes stem from one bidder having an actual or perceived advantage or disadvantage over the remaining bidders. This may occur when interested contractors call the engineer seeking advice and interpretations of the plans and specifications prior to bid opening. The engineer has no obligation to provide this information and, in fact, is well advised not to accept any calls. Most bid instructions require contractors to put requests for clarification in writing. Stick to this requirement and upon receiving the clarification requests, respond only through the issuance of addenda.

Contractors are only bound to perform to the contract and addenda. The generous offering of advice and direction prior to bidding may not only result in a bid protest but also cause disputes during construction when a subsequent decision of the engineer differs from a prebid opinion.

Similarly, during prebid conferences, answer all questions with reference to the contract documents. Try to control the natural impulse to answer "off the top of your head." If a question is not addressed in the documents, take it under advisement and issue an addenda.

Bidding procedures for public clients are usually creatures of statute. State statutes may require separate bidding for certain subcontract trades or contractor prequalification and certification for bidding certain types or classes of work. The bid documents must be consistent with these requirements. To avoid a bid protest because of faulty bid procedure, make sure the procedure used is consistent with local requirements and give the client the opportunity, in writing, to have the client's legal counsel review the bidding instructions for this purpose.

Whether public or private, the bid process is designed to award the contract to the lowest responsible and responsive bidder upon recommendation by the engineer. Hence, the engineer will be called on to make a recommendation as to contract award to his client. In order to do so, an evaluation of contractor responsibility and responsiveness is required. Evaluations as to responsibility and responsiveness are entirely different efforts.

A responsible bidder is a contractor with the ability to perform the project. This is measured by an evaluation of the contractor's experience, organization, skill, financial resources, availability (i.e., is the contractor spread too thin with other work?), integrity, and reputation. This evaluation is not an objective one, but calls for judgment after comprehensive research.

Some projects require an evaluation of contractor responsibility prior to bidding. Prequalifying contractors in this manner has some distinct advantages, including eliminating unqualified contractors from the bidding pool to promote fairer bids, promoting timely bid evaluation and award, and protecting the owner from unqualified low bidders. Whether pre-qualified or post-qualified, an evaluation of bidder responsibility should include the following:

1. **Require as part of the prequalification application, or bid submission, written evidence of financial resources, previous experience, and appropriate certifications of authority to perform the type of work bid.** Be specific and comprehensive as to what must be submitted. Review this information, keeping in mind it only tells half of the story.
2. **Check all certifications of qualifications.** Check state agencies for any disciplinary actions taken against the bidder. Check all references.
3. **Do an independent check on performance by calling colleagues and asking their experience with the bidders.** Identify and contact project owners not listed as references.
4. **Most importantly, summarize all investigations in writing.** There must be written evidence of the investigation to support whatever recommendation is ultimately made to the client.

A typical and basic contractor evaluation form is shown on Figure 8.

Contractor Evaluation Report

CLIENT CONTACT

PROJECT COMPANY

LOCATION

CONTRACTOR

ADDRESS PERSON

DESCRIPTION OF WORK

AMOUNT TIME FRAME DATE

RECORD OF PERFORMANCE

SCHEDULE OF WORK, MATERIAL, SUBCONTRACTORS, TIMELINESS

RATING

QUALITY OF WORK

RATING

MANAGEMENT QUALITY & RESPONSIVENESS, HOME OFFICE SUPPORT

RATING

AVAILABILITY OF EQUIPMENT AND SKILLED LABOR

RATING

OVERALL PERFORMANCE

RATING

REMARKS

RATING

DESCRIPTION OF WORK REVIEWED BY

RATING BASIS: 1 THRU 10 1– 4 UNSATISFACTORY 5–7 AVERAGE 8–10 GOOD TO EXCELLENT

Figure 8.

Engineers are reluctant to recommend rejection of a bidder for irresponsibility since bid protests are a near certain consequence. Such a determination is supported by courts so long as there is a reasonable basis for the determination. Any irresponsibility determination that was made arbitrarily or capriciously will be rejected by a court, leaving the owner-client responsible for damages to the disgruntled bidder. The key to showing a reasonable basis is a written summary of the bidder investigation and conclusions. This written summary and recommendation should be submitted to the client who must then make the determination whether to reject a bid based on irresponsibility.

This decision rests with the client and the engineer's role is limited to technical advisor.

Bid responsiveness is a more objective evaluation since it involves little more than confirming that the bid offer has been made in a form required by the bid instructions. Typically, this means that the contractor has properly filled out the bid form, acknowledged receipt of all addenda, attached proper certifications, and properly signed the bid. It is truly surprising that a concept as simple as bid responsiveness is so frequently made complex in the final rush and scurry before bid submission. Quotations are not properly written out, forms left blank, signatures are missed, math errors abound and a variety of other innocent, but nonresponsive, errors are commonly made in bids. All too frequently, such defects cause bid rejection, leading to bid protests.

Since the final moments before bid opening can be hectic, it is a good idea to have a detailed outline of even the most obvious of bid opening procedures. In this way, the bid opening procedure itself will not be cause for a bid protest. Figure 9 is an example of such a procedure which, of course, would need to be tailored to specific project requirements.

Generally, a bid defect that is a mere formality and does not affect the obligations of the parties entering the contract may be waived. This decision is not that of the engineer. The engineer's only responsibility is to identify the defect. Once identified, the owner should be informed, in writing, and advised that legal counsel needs to determine whether the defect can be waived or is fatal to the bid. In any event, a bid document should always reserve the right of the owner to reject all bids at his or her discretion and the right to waive any minor defect in the form of the bid that does not relate to price or performance of the project.

Because of financial restrictions and the uncertainty of the bidding environment, many bids provide for additive or deductive alternates. Alternate bidding is an effective way to control project price. When misused, it is also an effective way to manipulate an award to a particular bidder. In the public realm, improper alternate bidding may not only result in bid disputes, but is also illegal. The surest way to avoid even the appearance of bid manipulation by alternate bidding is to clearly prioritize the manner in which the alternates are applied against the base bid to determine the lowest bid for an award. This methodology should be defined in the instructions for bidders and must reflect the priority that the client-owner gives to the various portions of the project which would be added or deducted in the process. To avoid a later dispute with the client, ask for the client's preferences as to this priority in writing.

Unit price bidding can also generate later disputes. A common practice among contractors is to front load a unit price bid thereby getting paid a dispro-

Bid Opening Procedures

1. Call and remind Client of bid opening date and time.
2. Verify if Client has scheduled and set up a conference room for the opening.
3. Check with Client to make sure all key personnel are scheduled to be at opening.
4. Check verified mail list to insure all plan holders received addendum and bring list to bid opening.
5. Prepare copies of the bid tab for all in attendance.
6. Upon arrival, verify where bids are received, names of all Owner's representatives and their positions/titles, and who will actually read the bids aloud. Check to see if any mailed bids have been received.
7. Check the official time at the place of opening and start announcing the time at regular intervals.
8. Pass around attendance list with place to sign, representing who, telephone number and address.
9. At the time designated, state "Since it is now _____ o'clock local time, no further bids will be received for this project."
10. Introduce the representatives present.
11. Open each bid separately, checking execution, bid bonds, addendum acknowledgment, certifications and any other special bidding requirements.
12. Owner's representative should read the bid aloud while the engineer is keeping the bid tab.
13. Owner should then announce "The apparent low bidder is ________" and offer to read the low bidder's unit price items aloud if requested.
14. Thank all bidders for bidding, remind them to turn in their plans and specifications for refunds.
15. Tell everyone a certified bid tab will be mailed out to all bidders.
16. Review with the apparent low bidder, the next steps proceeding toward contract award.

Figure 9.

portionately higher price for early project work as compared to later project work. If the project should become troublesome mid-construction the engineer may find that too much money has been paid out, leaving an insufficient retainage to insure completion. To address this potential problem, the payment provisions of the contract documents should limit any periodic payment to an extension of the unit prices or an estimate of the fair and reasonable (and documented) value of the work completed to date, whichever is less.

Not only do contractors unbalance bids, but engineers in many instances inaccurately estimate quantities. Disputes then arise when the quantity originally estimated differs significantly from the unit price quantity in the bid documents. The inaccuracies in bid quantities are to be expected since many unit bid items are difficult, if not impossible, to predict prior to construction. A suggested solution to avoid disputes in this area is to set a percentage variance to the quantity estimate thereby allowing for a negotiated unit price change in the event an estimate proves grossly inaccurate. A 15% to 25% variation seems a reasonable limitation on usual unit price items. A higher or lower variation should be considered under unusual circumstances. The point is a clearly defined limit will avoid a dispute as to when a contractor is entitled to a change.

5.5 Preconstruction Meeting

The preconstruction meeting is the first opportunity for the engineer, owner, and contractor to develop a working relationship and start the project smoothly. It is usually the most friendly of all project meetings since the parties have yet to be presented with the inevitable construction problems. Use this meeting to define how the project will be managed. Establish a comprehensive agenda to discuss the contract's administrative requirements. Also, make sure all interested parties are in attendance. This would include not only the owner, contractor and engineer, but also representatives from the funding agency, power company, water department, sewer department, telephone company, fire department, police department, cable company, gas company, or dig safe as is appropriate. The preconstruction conference is the first and best opportunity to resolve confusion and avoid later misunderstandings. A suggested agenda is shown on Figure 10.

5.6 Substitutions

From the standpoint of professional responsibility, engineers have the obligation to evaluate and specify material and equipment that best respond to the project's needs. This traditional responsibility has come in conflict with the desire of many clients, especially public clients, that the specifications allow competition between equipment suppliers in order to promote cost competition and lower bids. Hence, engineers allow flexibility in the specifications through setting performance requirements only or allowing "equals" to the products specified.

Most engineers would favor proprietary specifications since the control of product selection rests with the engineer. There may be some pricing surcharge by a named supplier, under these circumstances, but the extent of any price surcharge is somewhat tempered by the knowledge that an excessive price will lead to being written out of future specifications.

Leaving product selection to the contractor through performance or or equal specifications may lead to some cost-savings usually reflecting a product inferior to what the engineer would have named. The contractor's principal incentive is to deliver materials and equipment that complies with the specifications at the lowest cost to the contractor. Superior products rarely meet this criteria. What the client saves in construction costs may be more than offset by the life cycle costs of products inferior to what the engineer would have named.

Pre-Construction Meeting Agenda

1. Submittal of executed bonds and insurance certificates.
2. Distribution of contract documents.
3. Submittal of list of subcontractors, list of products proposed for installation, schedule of values, and progress schedule.
4. Designation of responsible personnel.
5. Submittal of list of products proposed for substitution.
6. Procedures and processing of field decisions, submittals, substitutions, applications for payments, proposal requests, change orders, and contract closeout procedures.
7. Certified payroll requirements for Contractor and subcontractors.
8. Scheduling.
9. Use of premises by Owner and Contractor.
10. Owner's requirements and occupancy.
11. Construction facilities and controls provided by Owner.
12. Construction facilities and controls provided by Contractor.
13. Temporary utilities provided by Owner.
14. Field engineering.
15. Security and housekeeping procedures.
16. Payments to Contractor.
17. Procedures for testing.
18. Procedures for maintaining record documents.
19. Requirements for startup of equipment.
20. Inspection and acceptance of equipment put into service during construction period.
21. Substantial completion of work.
22. Final completion of work.

Figure 10.

In the case of proprietary specifications, the contractor has no right to substitute a named product. The engineer and owner have the right to insist on strict compliance. Keep in mind, however, that a substitution request may originate from the contractor's concern that the named product will not perform. Under these circumstances, the engineer should use the substitution request as an opportunity to recheck the plans and specifications to insure that the design is not somehow defective. Also, a contractor may present a substitution as a significant cost-savings to the owner. If such a cost-savings exists and the substitution proposed is acceptable (if not exactly equal), the engineer should advise the client, in writing, and let the client make the decision.

In the case of performance or "or equal" specifications, the contractor selects the materials and equipment and the engineer must then evaluate whether the materials and equipment meet the specifications or are "equal" to the products named in the specifications. In this sense, there is no substitution for a product, since no product has been preselected in the contract documents. In practice, however, the engineer is well aware of the *preferred* product upon which the design was based and goes about evaluating the contractor's proposal as a substitute. How should this evaluation take place?

One way is to shift the burden of investigating and evaluating the substitution onto the contractor by operation of the specifications. An example of this approach is shown in Figure 11. Another way is to require the contractor to make certain representations regarding the substitution as part of the substitution request as shown in Figure 12. An indirect approach that leaves control of the evaluation in the hands of the engineer is to require the contractor to compensate the engineer for the work involved in substitution evaluation. For instance,

> There shall be no increase in Contract Price for redesign due to substitution of products. The Contractor shall pay all reasonable fees of the Engineer in making an evaluation of the substitution request whether such request is ultimately approved or not.

Few contractors willingly abide by such provisions since most consider product selection a necessary right to transact business and remain competitive. Many, in fact, will lower bids in anticipation of having less costly substitute equipment and materials approved by the engineer during construction. Such approvals **should not** be made lightly since once approved, the engineer will be held as responsible as if the equipment had been named.

A substitution evaluation must consider the following:

1) **Performance.** Will the substitute perform as required by the specifications?

2) **Quality.** Is the substitute of near or equal quality in materials and fabrication?

3) **Durability.** Will the substitute last as long?

4) **Operation.** Will the substitute have similar operation and maintenance requirements?

The meaning of the term "or equal" should be set forth in the contract such that all parties have a basis from which to evaluate a product. For instance:

> An item shall be considered equal to the product identified in the specifications if (1) it is at least equal in quality, durability, appearance, strength and design; (2) it will perform at least equally the function imposed by the general design for the work being contracted for or the material being purchased, and (3) it conforms substantially, even with deviations, to the detailed requirements for the product in the said specifications.

Sample Substitutions Specifications

PART 1 GENERAL

A. Products Specified by Naming One or More Manufacturers with a Substitution Paragraph: Submit a request for substitution for any manufacturer not specifically named.

1.02 PRODUCTS LIST

A. At preconstruction conference, transmit three copies of a list of major products which are proposed for installation, including name of manufacturer.

B. Tabulate products by Specifications section number, title, and Article number.

C. For products specified only by reference standards, give manufacturer, trade name, model or catalog designation, and reference standards.

D. Engineer will not reply to this list of major products. All approvals, rejections and comments by the Engineer will be reserved until review of conforming equipment and material submittals and of formal request for substitution submittals.

1.03 LIMITATIONS ON SUBSTITUTIONS

A. Many sections of these specifications are based on specific manufacturer recommendations. Use of these recommendations does not and is not intended to exclude equal equipment of other manufacturers. The specifications serve only as a guide to minimum quality and performance.

1.04 REQUESTS FOR SUBSTITUTIONS

A. Submit separate request for each substitution. Document each request with complete data substantiating compliance of proposed substitution with requirements of Contract Documents.

B. Identify product by Specifications section and Article numbers. Provide manufacturer's name and address, trade name of product, and model or catalog number. List fabricators and suppliers as appropriate.

C. List similar projects using product, dates of installation, and names of Engineer and Owner.

D. Give itemized comparison of proposed substitution with specified product, listing variations, and reference to Specifications section and Article numbers.

E. Give quality and performance comparison between proposed substitution and the specified product.

F. Give cost data comparing proposed substitution with specified product, and amount of net change to Contract Sum.

G. List availability of maintenance services and replacement materials.

H. State effect of substitution on construction schedule, and changes required in other work or products.

1.05 CONTRACTOR REPRESENTATION

A. Submission of a request for substitution constitutes a representation that Contractor has investigated proposed product and has determined that it is equal to or superior in all respects to specified product.

B. Contractor will provide same warranty for substitution as for specified product.

C. Contractor will coordinate installation of accepted substitute, making such changes as may be required for Work to be complete in all respects.

Figure 11.

Sample Substitution Request Form

Date: ________________

1. Name of product to be substituted: ______________________________
2. Name of product requested as substitute: ______________________________
3. Specification Section reference: ______________________________
   Drawing Number reference: ______________________________
4. Preconstruction conference substitution list reference: ________________
5. Attach product data to this form.
6. List similar projects using this product:

	Project	Engineer
i.	______________________	______________________
ii.	______________________	______________________
iii.	______________________	______________________

7. Attach itemized comparison by Specification paragraph.
8. State effect of substitution on:
   i. Construction schedule: ______________________________
   ii. Project cost: ______________________________
   iii. Changes required in other work: ______________________________
9. Contractor representations
   i. Contractor has investigated proposed substitution and has determined that it is equal or superior to the product specified.
   ii. Contractor will provide same warranty and will coordinate installation of substitution through completion.
10. Certification:

Signed______________________

Title______________________

Company______________________

(Contractor)

Figure 12

Finally, as noted earlier, disputes arising from substitution requests can be avoided, in part, by designing the project to accommodate the range of suppliers from which the contractor may choose.

5.7 Shop Drawings Review

In principle, shop drawings provide the details of construction that are impracticable to present on plans and specifications for a number of reasons. For instance, the engineer is not responsible for and cannot predict the means and methods employed by contractors to meet project requirements. Shop drawings identify these details of installation. In projects having non-proprietary specifications, the engineer does not know which manufacturers will be used and consequently cannot design for a particular product's requirements. The shop drawing bridges this gap between the engineer's design and the contractor's implementation by identifying the products to be used and the details of construction.

Shop drawings are typically forwarded from the contractor to engineer for review and approval. The shop drawings, or more generically, *the submittal*, consists of installation drawings, product data, samples, if appropriate, and a contractor's representation as to compliance with the contract requirements. A typical shop drawing specification is shown in Figure 13.

Sample Shop Drawings Specifications

PART 1 GENERAL

1.01 SHOP DRAWINGS

A. Title each drawing with Project name and number; identify each element of drawings by reference to sheet number, detail and schedule, of Contract Documents.

B. Identify field dimensions; show relation to adjacent or critical features of Work or products.

1.03 PRODUCT DATA

A. Submit only pages which are pertinent; mark each copy of standard printed data to identify pertinent products, referenced to Specification Section and Article number. Show reference standards, performance characteristics, and capacities; wiring and piping diagrams and controls; component parts; materials of construction; finishes, dimensions; and required clearances. Provide Material Safety Data Sheets required by OSHA for all chemicals to be supplied under this contract.

B. Modify manufacturer's standard schematic drawings and diagrams to supplement standard information and to provide information specifically applicable to the Work. Delete information not applicable.

C. Provide manufacturer's preparation, assembly, and installation instructions.

D. If similar or identical submittal material can be submitted under more than one specification section, the Contractor shall make separate clearly identified submittals for each specification section. Submittal materials for any specifications section shall be complete for that section; partial submittals are unacceptable.

1.04 SAMPLES

A. Submit full range of manufacturer's standard finishes except when more restrictive requirements are specified, indicating colors, textures, and patterns, for Engineer selection.

B. Submit samples to illustrate functional characteristics of products, including parts and attachments.

C. Approved samples which may be used in the Work are indicated in the specification section.

D. Label each sample with identification required for transmittal letter.
E. Provide field samples of finishes at Project, at location acceptable to Engineer, as required by individual specifications section. Install each sample complete and finished. Acceptable finishes in place may be retained in completed work.

1.05 CONTRACTOR REVIEW

A. Review submittals prior to transmittal; determine and verify field measurements, field construction criteria, manufacturer's catalog numbers, and conformance of submittal with requirements of Contract Documents.
B. Coordinate submittals with requirements of Work and of Contract Documents.
C. Sign and date each sheet of shop drawings and product data, and each sample label to certify compliance with requirements of Contract Documents. Notify Engineer in writing at time of submittal, of any deviations from requirements of Contract Documents.
D. Do not fabricate or ship products or begin work which requires submittals until return of submittal that Engineer has reviewed and approved.
E. Shop drawings, product data and sample submittals shall be made in a timely and logical fashion taking into account work scheduling and job progress.

Figure 13.

An approved shop drawing is generally a prerequisite to releasing the contractor to have the product or material fabricated. The engineer must then review the shop drawing for compliance with the plans and specifications as well as the "design intent" of the project. The shop drawing should be checked against each provision of the specification, then checked against the specifications and shop drawings for interrelated equipment and materials. A sample shop drawing review procedure is shown on Figure 14.

Unfortunately, the shop drawing process does not work in practice as well as

Sample Shop Drawing Review Procedure

A. SHOP DRAWING PROCESSING
1. Shop drawings are normally submitted by the Contractor to the Project Manager.
2. The submittals should be logged in, numbered and dated.
3. The Project Manager will distribute them in three ways:
    i. Design Project Engineer—submittals requiring design engineer review.
    ii. Project Manager—submittals not requiring design engineer review.
    iii. Outside consultants—submittals for their design responsibility.
4. In each of the above cases, a review form should be attached to the submittal with any appropriate comments. This must be completed and returned by the specified date. Timely review is very important to eliminate a possible delay claim by the Contractor.
5. A copy of the completed review form should be sent to the Design Project Engineer.
6. The Project Manager will then return submittals to the Contractor.

B. KEY PERSONNEL FUNCTIONS

1. Administrative Assistant
   - Receive submittals from contractors.
   - Establish and maintain a numbering and filing system, with all submittals dated.
   - Maintain a "submittal log" for tracking the "in-house" review process.
   - Forward submittals and review form to Project Manager, Design Project Engineer, or outside consultant, as appropriate.
2. Project Manager
   - Assure that contractor submits shop drawings in accordance with an established schedule and in the required format.
   - Assure overall functioning of "in-house" review process.
   - Review shop drawings not requiring external assistance.
   - Establish a review schedule with Design Project Engineer.
   - Maintain control of shop drawings review budget.
3. Design Project Engineer
   - Review shop drawings in accordance with schedule established jointly with Project Manager.
   - Maintain control of shop drawing review budget.
   - Return completed submittal packages to Administrative Assistant or Project Manager as appropriate.
   - Coordinate shop drawing reviews with outside consultants.
4. Outside Consultant
   - Establish review schedule with Project Manager.
   - Conduct review in accordance with schedule and return submittal to Project Manager or Administrative Assistant.

C. CHECKLIST FOR VERIFYING SHOP DRAWINGS

1. Specifications
   a. Type
   b. Size (dimensions, volume, area, weight, etc.)
   c. Capacity (gpd, etc) at specified loading
   d. Design loads (structural, pressure, temperature)
   e. Type of service
   f. Materials
   g. Codes—existing codes/changes in codes
   h. Operating conditions (tolerance/ fluctuations)
   i. Noise load
   j. Accessories - spare parts
   k. Warranty
   l. Manufacturer training
2. Connections
   a. Electrical
   b. Piping connections
3. Motors
   a. Proper voltage
   b. Size/type/speed
   c. Service factor
   d. Explosion-proof
   e. Motor starter
   f. Controls
   g. Sheaves
   h. Guards
   i. Cooling requirements
4. Installation
   a. Physical fit
   b. Storage requirements
   c. Access
   d. Vibration isolation
   e. Equipment supports
5. Maintenance
   a. Start-up service
   b. Service agreement with owner
   c. Access for maintenance

Figure 14.

in principle. Contractors view the shop drawing process as an administrative headache that inevitably holds up job progress. Literature from manufacturers and drawings from fabricators work their way through subcontractors and the general contractor to the engineer unreviewed and unorganized. Instead of spending the time in submittal presentation, many contractors find it more cost-effective to repeatedly submit and resubmit whatever is submitted to them in the hope that the engineer will be liberal with approvals.

The engineer, inundated with non-responsive submittals, is pressured to process these submittals as soon as possible. Repeated resubmittals not only hold up job progress but also put added pressure on the shop drawing review budget. Also, some engineers buried in paper, faced with the tedious task of checking shop drawing details and concerned with budgetary constraints, may rely on the contractor's representations as to submittal compliance. Not surprisingly, such reliance may not be in the best interest of the project.

Contractually shifting the responsibility for shop drawing review to the contractor is a common response. For example:

> Contractor shall check and verify all field measurements and shall be responsible for the prompt submission of all shop and working drawings so that there shall be no delay in the work.

Just as common and more problematic, is the practice of providing contractors with approvals that at any time can convert to rejections. For example,

> Contractor shall be responsible for the consequences of revocation of the Engineer's approval because of the Contractor's failure to submit a complete submittal and/or to identify portions of the submittal that does not conform to the specifications.
>
> Regardless of corrections made in or approval given to such drawings by the Engineer, the Contractor will nevertheless be responsible for the accuracy of such drawings and for their conformity to the Contract Documents. The Contractor shall notify the Engineer in writing of any deviations at the time he furnishes such drawings.

Some engineers provide contractors with approvals limited to "design concept." A shop drawing stamp may read

> APPROVED..................................
> APPROVED AS CORRECTED......................
> REVISE AND RESUBMIT........................
> NOT APPROVED...............................
> Approval is only for conformance with the design concept of the Project and compliance with the information given in the Contract Documents. Contractor is responsible for dimensions to be confirmed

and correlated at the job site; for information that pertains solely to the fabrication processes or to techniques of construction; and for coordination of the work of all trades.

ENGINEER

By__________________________ Date____________

Presumably, "design concept" is whatever the engineer wants it to mean at the time a problem arises.

Such risk shifting or responsibility avoiding approaches to shop drawing review provide a false sense of security and foster later disputes. Engineers are held responsible for the manner and thoroughness in which shop drawings are processed. As such, the engineer is well advised to:

1. **Restrict shop drawings to necessary equipment and materials.** Engineers have the tendency to require shop drawing submittals for every conceivable product on the project. This is simply unnecessary. If the plans and specifications are sufficiently detailed, there is no need for a shop drawing. Shop drawings, once approved, can act to change the contract from what was intended to an unintended inferior or unacceptable alternative. There is no sense in taking the chance of an improperly reviewed shop drawing unintentionally modifying a good specification. Also, unnecessary submittals promote submittal volume, taking away precious engineering time that would be better utilized in focusing on key submittals.
2. **Practice preemptory submittal rejection.** Submittals that are carelessly prepared and unorganized should be immediately rejected. Engineers are normally not involved in submittal preparation, only submittal review. Speedy rejection of improperly presented submittals will greatly increase the likelihood that properly prepared and more importantly, easily reviewed submittals will be forthcoming.
3. **Timely and competent response.** The contractor's performance schedule is tied directly to submittal approval. Materials and equipment can't be delivered and installed without approvals to fabricate. Despite the tedious nature of the work, the shop drawing review process should be a priority service and all responses must be timely. Unreasonably long periods of review can cause delays for which the engineer could be held responsible. Similarly, the shop drawing process must employ experienced engineers who can readily spot non-compliance.
4. **Adequate Budgeting.** The owner-client must be educated as to the importance of "necessary" shop drawing review. An adequate budget, with contingencies should be established for this purpose.

The 114 dead and 85 injured as a result of the collapse of the walkways of the Kansas City Hyatt Regency Hotel would have avoided their fate had the shop drawings on that project been handled differently. A closer examination of the structural connection depicted on the shop drawings might have revealed that the redesign of the walkway connections had doubled the load beyond the capacity of the connections. Certainly, the Kansas City Hyatt was a tragedy of severe and hopefully unique proportions. The point, however, is the Kansas City Hyatt's failure as well as the vast majority of less significant shop drawing related problems, can be avoided if submittals are properly managed and reviewed.

5.8 Scheduling of Construction Activities

The scheduling of construction activities is generally left to the general contractor. The engineer, however, plays a key role in both setting the overall project schedule, monitoring compliance with the schedule and, when appropriate, phasing construction.

The overall project schedule is set when the contract time of completion is determined. In order to determine how long a project will take, the engineer must predict how a contractor will approach the project work taking into account seasonal shut down periods. Usually, the time set for performance is generous. Occasionally, due to external time pressures, such as regulatory agency orders, the time of performance is limited and the engineer must determine whether such time limitations can be achieved. Obviously, requiring performance with an impossible time frame will not be an enforceable contract time and instead indicate a defect in the contract documents.

Once the overall project time is set, incentives to complete are usually identified. These could be either bonus provisions for early completion or liquidated damages provisions for late completion. Liquidated damages must be reasonable estimates of actual loss in the event of late completion and not simply penalties. As noted earlier, prepare written documentation as to the determination of a liquidated damage rate.

Monitoring compliance of a project schedule is a critical function in that failure to meet key milestone dates is the first indication of a project in trouble. The easiest way to monitor a general contractors compliance with the progress schedule, is to specify the type of schedule to be maintained and the methodology of usage. For instance, Figure 15 is a typical specification on progress schedules. It identifies the type of schedule to be maintained, frequency of updates and degree of detail. All too often, engineers fail to enforce periodic schedule submittals. Contractors are reluctant to focus on this and other administrative-type provisions of the contract because paperwork is not viewed as

Sample Progress Schedules Specifications

PART 1 GENERAL

1.01 FORMAT

A. Prepare Schedules as a horizontal bar chart with separate bar for each major portion of Work or operation, identifying first work day of each week and identifying each portion of the work that is critical to timely project completion. All project scheduling shall include and be prepared using critical path method analysis.

B. Sequence of Listings: The chronological order of the start of each item of Work.

C. Scale and Spacing: To provide space for notations and revisions.

D. Sheet Size: Multiple of 22 x 34 inches.

1.02 CONTENT

A. Show complete sequence of construction by activity, with dates for beginning and completion of each element of construction.

B. Identify each item by major specification section number.

C. Identify work of separate stages or separate floors, and other logically grouped activities.

D. Show accumulated percentage of completion of each item, and total percentage of Work completed, as of the last day of each month.

E. Provide separate schedule of submittal dates for shop drawings, product data, and samples.

1.03 REVISIONS TO SCHEDULES

A. Indicate progress of each activity to date of submittal, and projected completion date of each activity.

B. Identify activities modified since previous submittal, major changes in scope, and other identifiable changes.

C. Provide narrative report to define problem areas, anticipated delays, and impact on Schedule. Report corrective action taken, or proposed, and its effect.

1.04 SUBMITTALS

A. Submit initial Schedules at the preconstruction conference. After review, resubmit required revised data within ten days.

B. Submit revised Progress Schedules for review and approval, with each Application for Payment.

Figure 15

an efficient means of advancing the project to completion. Nevertheless, periodic updates of schedules are key indicators of project problems and later disputes. One sure way of obtaining timely updates is to require schedule submittals as part of the periodic applications for payment. Without a schedule, the application is simply not complete and cannot be paid.

Once schedules are submitted, they can't be ignored. With information comes responsibility and the engineer must assess and identify potential project delays. More importantly, these delays need be reported to the client and addressed with the contractor. If time extensions are warranted, address an extension **before** the milestone date is missed.

It is not uncommon for contractors to manipulate the project schedule by claiming early completion or float sequestration. Early completion involves the submittal of an initial schedule indicating project finish prior to the contract completion time. The contractor claims the right to finish early and when the project actually completes on time or late, the contractor claims delay damages from the early finish date. For example, the contractor may prepare the early finish schedule assuming an unreasonably short turnaround time of shop drawings. Inevitably, the engineer will fail to meet the contractors assumed turn around time, which will result in a delay claim.

Sequestering of float involves project scheduling of *all* items as critical. Hence the delay of *any* task causes a delay to the entire project, thereby justifying a time extension.

To address the early completion problem and avoid the claim, the specifications should include a clause that limits delay damages to after the contract completion date. For instance,

> The Contractor is not entitled to recover delay damages for the period of time between the contractor's scheduled completion date and the contract completion date.

Hence, under this provision there is no advantage to the contractor in scheduling the project in less than the full contract completion time.

Similarly, float manipulation can be addressed in the specifications. For example,

> Contractors shall not engage in float manipulations which have the net effect of sequestering float time. Examples of scheduling techniques disallowed under this provision include extending time duration estimates.

This anti-float manipulation clause effectively requires the contractor to establish a unique critical path that all parties to the project can use to gauge progress.

Certain projects require the engineer to become involved in the contractor's schedule of its work. Rehabilitation projects on operating facilities is an example of a situation where the engineer must determine the sequence of the contractors activities to assure that the facility remains operational. Typically, phases of construction are defined with milestone dates. Here the engineer must make the effort of careful evaluation of the proper sequence of work since contractors, having little knowledge of the operational requirements of the owner's facilities, cannot realistically identify a proper work sequence. This is one area where the engineer can be held responsible not only by the contractor for delays caused by improper sequencing but also by the owner-client for interruptions in facility operations caused by improper sequencing.

5.9 Construction Inspection

Whether referred to as monitoring, observation, periodic site visitation, or resident engineer representation, a key service provided to owners by engineers is construction inspection. The engineer becomes the owner's "eyes and ears" on the project, monitoring both work progress and compliance. Theoretically, the engineer's role is that of an observer, since the contractor is responsible for the means and methods of construction. In practice, the engineer's role on the project is much more pervasive.

The "resident project representative" or simple the resident engineer, is the key person involved in construction inspection. The resident engineer is the point person on site representing the engineering firm and the owner. The contractor's superintendent presents the endless array of daily construction concerns to the resident engineer who provides direction, or at a minimum, referral to the engineer's office staff. The superintendent and resident engineer, for all practical purposes, drive the project towards completion. The success or failure of any construction project depends on the competence of the contractor's superintendent and the resident engineer's.

There is no question that close cooperation between the resident engineer and superintendent promotes construction efficiency. As site problems or questions arise, an immediate and fair set of instructions and direction from the resident engineer, as the owner's representative, binds the owner to whatever directions and instructions were provided. The resident engineer may change the contract to the extent that any direction differs from what the contract would have otherwise required. For this to occur, the resident engineer is then considered to have authority.

Authority can be either expressed, implied or apparent. Expressed authority is that authority given to the resident engineer by some written or oral directive. For instance, the resident engineer is usually given expressed authority to review payment applications. Implied authority is authority vested in the resident engineer by the circumstances, without specific expression. For instance, the very nature of the resident engineer's position would indicate the resident engineer has the authority to observe the contractor's work and record progress. Apparent authority is authority which is apparent to third parties by the behavior of the resident engineer. For instance, if the resident engineer has consistently changed the contract by relieving the contractor from certain performance, and no objection was made by the engineer or owner, the resident engineer is considered to have authority to change the contract.

As one can imagine, how much authority a resident engineer has or exerts over the contractor's performance depends, in large part, on how proactive the resident engineer chooses to be. To resolve the potential problem of a runaway resident engineer, construction contracts can include provisions that define the resident engineer's authority or limit that authority. For instance:

Resident Project Representative:

1. Shall not authorize any deviation from the Contract Documents or substitutions of materials or equipment.
2. Shall not undertake any of the responsibility of contractor, subcontractors or contractor's superintendent.
3. Shall not advise on, issue directions or assume control over any aspect of the means, methods, techniques, sequences or procedures of construction.
4. Shall not advise on, issue directions regarding or assume control over safety precautions and programs in connection with the Work.

A contractor can't claim a resident engineer had authority to take an action when the contract expressly limited that authority. This, of course, doesn't prevent many residents engineers from attempting to exercise such authority in the face of a contractual limitation.

Given the importance of the residential engineer, it is surprising that many engineering firms consider the position as entry level and offer very little training. In many cases, resident engineers learn on the job at the expense of the clients. In some instances, the resident engineer lacks an appreciation of the limitations of his or her authority and the sanctity of the design. All the time and effort expended in the details of design are for naught if the resident engineer decides, in the field, that there is a "better way to do things." A resident engineer that takes over construction and, in effect, becomes the contractor's quasi-superintendent is bound to create serious liability for the engineering firm and the owner even when the project appears to be moving along smoothly.

A resident engineer who takes a confrontational stance to the contractor's every inquiry is just as problematic as the overly cooperative residential engineer. Nothing gets resolved in the field and every inquiry is transmitted to the office, taking unnecessary time and expense for resolution.

The overly cooperative or confrontational resident engineer situation may stem from the project manager's lack of field experience. Project managers are typically licensed engineers who have focused their experience on technical and design issues. Few project managers have extensive construction experience, yet it is the project manager who is responsible for determining whether the contractor's work conforms to the design. In order for a resident engineer to perform properly in the field, he or she must rely on timely and competent decisions by the project manager.

To this end, the following suggestions are offered:

1. **All project managers should be required to have field experience.** Design experience alone is not enough to manage a construction project. As a further advantage, project managers having construction means and methods experience will perform design work that is more "buildable."
2. **Resident engineers should receive proper training.** Although an engineering degree and license may not be necessary, resident engineers must be trained inspectors and knowledgeable as to construction means and methods. The resident engineer must also understand the limitations of the position and at all times protect the integrity of the design against the expediency of construction in all field decisions.

The resident engineer's principal role is that of observer and recorder of construction progress. Key to this role is the documentation prepared by the resident engineer, especially the field reports. Detailed field reports are **the** most useful construction documents in responding to disputes. With detailed field reports, the events that gave rise to a claim can be determined and subsequent costs relating to a claim can be evaluated. With poor or missing field reports, a dispute can degenerate into the contractor's word against the resident engineer's word, both of which are subject to competing interests and the inaccuracies of human memory. The answer lies in the preparation of contemporaneous accurate and comprehensive field reports. A suggested procedure for daily field reports is shown on Figure 16.

Assuming a field competent project manager and position-sensitive resident engineer, one of the most effective opportunities to resolve disputes are periodic project meetings. At these meetings, the project manager or engineer can address field problems with the contractor and resident engineer on the site. It is important that these progress meetings be structured in order to cover as many problems or potential problems as possible. It is just as important that these meetings result in decisions. Taking problems under consideration over extended periods of time during construction is sure to result in disputes and delay claims. A suggested outline for a progress meeting is shown on Figure 17.

5.10 Change Order/Claim Evaluation

There are three basic types of change orders:

1. owner generated;
2. contemplated by the contract; and
3. contractor generated.

An owner-generated change order occurs when the owner elects to increase

Daily Feild Reports
Procedure

General Rules:

1. Written at the end of the day.
2. Printed or handwritten in clean, neat, readable manner.
3. Use a daily diary as supporting document.

The daily field report should cover the following:

1. Project Identification and location.
2. Date—Month, day, year and number of days since project commencement.
3. Report Number—Should correspond, if possible, with calendar days since project commencement.

Report weekends whether worked or not.

4. Weather and Temperature—Record both A.M. and P.M. conditions with times, if possible.
5. Calendar Days—Contracted, Consumed, Remaining—Coordinate with contract time and reference to Report No., if possible.
6. Subcontractors—List only if on-site that day.
7. Classification of Labor.
8. Classification of Equipment—Should reflect hourly equipment use schedule. List all equipment whether idle or not.
9. Classification of Work.
10. Visitors.
11. Description of Project Activities
    a. Typically begin with work by General Contractor and general observations.
    b. Clarify General Contractor's division of labor, if necessary.
    c. Specifically describe work performed, how many men and how long.
    d. Note names or workmen, if possible.
    e. Identify equipment used and for what purpose.
    f. List materials used.
    g. Perform same for each Subcontractor.
    h. Note effect weather may have had on project, if any.
    i. Note work performed by inspection staff.
    j. Identify visitors to site and reasons on-site.
    k. Identify major deliveries to site and log into estimate book.
    l. Coordinate photo log entries to report. (Always back up key observations with photographs.)
    m. Identify any tests conducted.
    n. Note correspondence issued or important correspondence received.
    o. Note contractors, subcontractors and inspection staff work hours.
    p. Cross-reference to diary; check off all entries that have been covered in daily report.
12. Sign each page upon completion.
13. Number each page and record total.

Figure 16.

Progress Meetings

A. Engineer will schedule and administer Project meetings throughout progress of the Work at a maximum of monthly intervals.
B. Engineer will make physical arrangements for meetings, prepare agenda, notify Contractor as to representatives that should attend, preside at meetings, record minutes, and distribute copies of the minutes.
C. Agenda:
   1. Review of Work progress
   2. Status of progress
   3. Schedule and adjustments thereto
   4. Delivery schedules
   5. Submittals
   6. Pending changes and substitutions
   7. Other items affecting progress of Work

Figure 17

or decrease the scope of the contractor's work. In this event there is no dispute that a change has occurred and the engineer's obligation is simply to evaluate the value of the change and adjust the contract price and/or time of performance accordingly.

A change order contemplated by the contract is usually a mechanical application of quantities and price. The most common example of this type of change order is the balancing change order. A balancing change order is used to reconcile actual versus estimated bid quantities. The engineer's obligation is simply to compute the actual quantities of work and apply such quantities against the price set forth in the unit bids.

Disputes can arise as to the evaluation of a fair price or calculation of unit bid quantities. These disputes, however, are usually resolved using project documentation. For instance, invoices, receipts, and standard pricing data address a dispute as to the fair value of work. Similarly, detailed survey information and quantity takeoffs address disputes over balancing change orders.

Contractor generated change orders, however, present an entirely different set of considerations.

A contractor-generated change order is a claim for additional payment and/or time caused by a change in the scope of the work. In order to determine whether a contractor is entitled to a change in the contract price or time, the engineer must evaluate both the substantive and procedural aspects of the change order request. The substantive issue to be evaluated is whether the contract gives the contractor a right to change the terms of the contract. Usually the general conditions of the construction contract will specify which events give rise to such a right.

The most common events that entitle a contractor to a change price or time of performance are (1) orders of the engineer, (2) sub-surface conditions, (3) concealed physical conditions, (4) unknown physical conditions of an unusual nature, (5) contract interpretations by the engineer, (6) stop work orders from the owner or engineer, (7) failure to be paid by the owner, (8) termination of the contract by the owner, (9) owner suspension of the work, (10) labor disputes, (11) fire or casualties, (12) unusual delay in deliveries, (13) engineer's failure to make timely interpretations, (14) adverse weather conditions, (15) ambiguous specifications provisions, (16) conflicts and inconsistencies among the various contract documents, and (17) defects in plans and specifications.

Occasionally the substantive right to a change in the contract is limited only to time of performance. For instance, in most standard contracts, unusually severe weather conditions allow the contract time of performance to be extended but does not entitle the contractor to a change in contract price.

Hence, in evaluating a change order, the engineer must first look to the terms of the contract and identify what, if any, contract term entitles the contractor to a change in the contract. Once the engineer identifies such a contract provision, the engineer must determine whether the contractor has met the procedural requirements of the contract. In other words, the engineer must determine whether the contractor is procedurally entitled to a contract change.

The most important procedural requirement of a construction contract with respect to claims is the notice provision. In most instances, when notice requirements are not followed the contractor waives the right to assert a claim. Hence, under those circumstances, and even if the contractor establishes that the contract scope has changed, no modification to either payment terms or time of performance is required.

The notice provision is based on the principle that if the owner or engineer knew that the contractor was going to claim that certain work was beyond the scope of the contract, then the owner and the engineer would have had the opportunity to take actions to avoid the claim or, at a minimum, monitor the magnitude of the work of the claim. Then, when the contractor submits its costs, the engineer would have an adequate base of information to evaluate the reasonableness of that cost. If the contractor failed to provide any notice, the engineer may not have monitored that aspect of the work and would be unable to reasonably evaluate the later claim.

The failure of the contractor to give proper notice is the most common first response used by engineers in rejecting contractors' claims. This response, however, should not be made without first considering whether the engineer had actual notice of the claim despite the contractor's failure to provide formal written notice. For example, a common construction project scenario is for the engineer to be aware that the contractor is disgruntled over performing certain work and intends to file a claim. The engineer then observes the contractor performing this work and the contractor fails to follow the formalities of the notice procedure thinking, "What's the point? The engineer already knows this situation." When the contractor finally completes the work and submits the claim, the engineer then takes the inequitable position that the contractor failed to follow the notice formalities and therefore the claim is rejected. Courts find ways around inequitable results and, in this example, the engineer's rejection is clearly unfair.

It is common practice for the engineer to administer the construction project inconsistently with respect to notice. For example, those claims that the engineer agrees with are commonly executed as change orders despite the fact that administrative formalities are not satisfied. This type of inconsistent behavior on the part of the engineer leads to the argument that the engineer waived the formalities of claim presentation, including notice, by failing to consistently enforce these formalities throughout the project. Under these circumstances, if the engineer had actual knowledge that the contractor was performing work the contractor believed was outside the scope of the contract, the engineer would be hard pressed to claim lack of notice despite the fact that the contractor may not have met the detailed formalities of notice.

The evaluation of change order requests presents more than just a challenge to evaluate substantive and procedural entitlement. It is common practice for the engineer to be identified as the interpreter of the contract documents. Also most construction contracts obligate the engineer to make such interpretations in an unbiased fashion. Clearly if the claim is unsubstantiated or if the claim identifies an obvious defect in the plans and specifications, the engineers decision is easily formulated. The vast majority of change order requests, however, fall in the gray area in which the contractor's interpretation supporting a change is just as reasonable as an interpretation that would deny a change. This is one area in which the engineer faces an apparent unresolved conflict of interest.

Given the fact that the engineer works for the owner, the engineer is naturally reluctant to explain to the client why the project will cost more money or take more time than anticipated due to a claim that the engineer's plans and specifications are defective. Instead, the engineer may favor a contract interpretation that denies the contractor's request and supports the original design. Any denial, if unsupported, can lead to serious liability on behalf of both the engineer and the owner. The key then is proper support for any decision and to that end the following recommendations are presented.

1. **Request clear and comprehensive documentation by the contractor.** Most contractors will provide a minimum of information from which the engineer is expected to make an evaluation and determination. The contract documents should set forth a clear description of the backup documentation necessary to support any claim for a change in scope or time of performance. Insist that the contractor comply with the requirements relating to supporting documentation.
2. **Administer changes to the construction contract in a consistent manner.** Do not periodically waive requirements for notice only to insist that such requirements be complied with when it suits the purpose of denying a claim.
3. **Clearly document claim evaluation.** Demonstrate that every claim was comprehensively and fairly evaluated. Prepare a clear record describing all efforts made in evaluating the claim.
4. **Work with the client.** Bring the owner into the process of claim evaluation as soon as possible. It is imperative that an owner understand that a set of plans and specifications cannot be perfected and that defects may entitle the contractor to additional payment and/or time of performance. The owner should be made to understand that such defects are not an indication of faulty engineering services.

5.11 Payment Certification

The administration of partial progress payments during the course of construction begins and sometimes ends with the resident engineer. Typically, the superintendent and resident engineer meet to determine the quantum of work performed and then the resident engineer applies that quantum, whether field measured or estimated, as percentage completed, against the schedule of values to arrive at a payment amount. As with so many other aspects of the construction process, the "golden rule" or "he who has the gold makes the rules," applies with partial payment administration. In this case, the resident engineer controls the gold and most contractors are wise enough to recognize this fact.

Accordingly, most superintendents will agree to all but the most severely inaccurate estimates by the resident engineer knowing full well that a partial payment application submitted in agreement with the resident engineer can be expected to be processed and paid promptly. The life blood of any contractor's business is prompt progress payments.

Conversely, disputed pay estimates require decisions of the project manager. The project manager who may rarely be at the site must rely on the resident engineer. Given such reliance, it is a rare payment application that is adjusted upward from the resident engineer's estimate by the project manager at the request of the contractor.

In response to this situation, contractors tend to overestimate task completion in anticipation of a later "negotiated" reduction. Also contractors delay payment to subcontractors as long as possible and rarely pay any subcontractor prior to payment being made for the subcontractor's work by the owner. Finally contractors tend to manipulate the schedule of values to disproportionately "qualify" for greater payments earlier in the project. All of these techniques cause later disputes.

To avoid payment-related disputes, the engineer should approach progress payments with the following in mind:

1. **The contractor's primary interest is to get paid as much as possible and as fast as possible.** This is simply business. Few contractors can finance a project and the absence of timely payments is sure to disrupt project progress. Hence, the engineer should make payment application review and administration a top priority.
2. **The owner should be advised and reminded of the importance of timely payment.** This is especially true for public owners, many of whom have become accustomed to long lead times between pay application approval and actual payment. The owner should be reminded that the contractor usually has specific contractual remedies in the event of late payment including the right to stop work.
3. **Carefully evaluate the contractor's schedule of values.** This schedule of values should be balanced and reasonably estimate the value of work as performed. Look for and correct any "front-end loading" of the schedule of values prior to commencement of construction. This will avoid later disputes when extending agreed upon quantities against price.
4. **Maintain complete and accurate records of project progress.** The resident engineer should be able to **accurately** determine project completion through field reports and indisputably demonstrate to the contractor this measure of project completion. Again, accurate field reports are key to avoiding a later dispute in this regard.

The engineer should be aware that the administration of payment applications effects subcontractors, sureties and funding agencies as well as the contractor and owner. It is not uncommon for subcontractors to complain to the engineer that they are not being paid. Although, the engineer has no obligation to assure payment to the subcontractors, the engineer is normally authorized to recommend retainage of funds to cover subcontractor claims of nonpayment. Also, the contractor is normally required to attest to payment of subcontractors as part of the payment application, or, in the case of private projects, submit partial lien waivers with each payment application. The engineer should make sure that these documents, as well as any other documents, are properly submitted.

With respect to sureties, the engineer and owner can compromise bond protection by overpaying a contractor who later defaults on contract obligations. In the event of a default, the surety has rights to any outstanding payments or retainage due its principal, the contractor. To the extent the surety's position in this regard has been compromised, the surety may raise overpayment as a defense to a bond claim. Again, a fair and accurate estimate of project progress avoids this problem.

Finally, payment application administration effects funding agencies. Funding agencies rely on the engineer's certification of a payment application and estimates of project completion to release funds to the owner for use in financing construction. Improper or negligent approvals of payment certifications could lead to overpayment of funds that breach funding agreements. It deserves repeating, a fair and accurate estimate of project progress, applied against a fair and accurate schedule of values results in proper payment to the contractor and avoids disputes with the contractor and others.

5.12 Safety Considerations

In most construction contracts as well as professional services agreements, there are provisions that expressly disclaim the engineer's obligations with respect to the contractor's means and methods and site safety. These contracts also include companion provisions that expressly state that it is the contractor who is exclusively responsible for the means and methods of construction and for all safety aspects of construction.

Despite these clearly defined limitations and responsibilities, resident engineers occasionally assume site responsibility during inspection. With good intentions, a resident engineer may direct the contractors' workmen in a manner that the resident engineer views as enhancing site safety. This is a clear expansion of the engineer's authority and if a site safety accident were to occur, the engineering firm could be held responsible.

Safety responsibility has also been extended to the engineer when the engineer fails to use the authority vested in the engineer to stop the work (if such authority is provided in the contract documents). A typical scenario is where the resident engineer views what is obviously a hazardous work practice, reports the same to the project manager, and the project manager elects to do nothing. In these circumstances, the engineer has been held responsible for the injuries that could have been avoided had the engineer exercised the authority to stop work.

As a consequence to this potential liability, few contracts give the engineer the right to stop work. Usually this right is exclusively vested with the owner.

One must keep in mind that the worker's compensation system promotes claims against non-employer parties. With most safety accidents, the contractor is at fault by providing an unsafe workplace. Worker's compensation laws, however, protect the contractor from being sued since the worker forgoes the right to bring suit in return for prompt insurance payments. The payout from the insurance program can be significantly less than the actual cost of the injuries sustained by the worker. Hence, the worker will search for a potential defendant that is not protected by the worker's compensation statutes. The engineer and, of course, the engineer's errors and omissions insurance company are obvious targets for this search. Although there has been some movement to include the engineer with the contractor as parties protected by worker's compensation laws, the general rule is that the engineer is not so protected.

Clearly a resident engineer cannot ignore a serious safety hazard that comes to his or her attention during the progress of the work. Regardless of the details of the contract language and the limitations of authority, the resident engineer will naturally react to any situation that could cause serious injury. It is the manner in which the resident engineer reacts that gives rise to liability. To avoid this liability, it is suggested that the resident engineer do the following in the face of a safety hazard.

1. **Document the safety hazard in a field report.** Include the details of the unsafe practice in the field report.
2. **Issue a field memo to the contractor's superintendent reporting the unsafe condition.** Remind the superintendent that it is the superintendent's responsibility to maintain a safe workplace. Do not assume the superintendent's responsibility and direct the work. Instead, notify the superintendent and let the superintendent take corrective action.
3. **Report the matter to the project manager in the office as well as the owner.** Since the owner may have the ultimate authority to stop work in response to a hazardous condition, the owner must be informed of dangerous site conditions promptly. This notification is best made by the project manager after discussion with the resident engineer.

The above recommendations may not be as expedient as simply directing the work to be performed in a different manner. It is, however, in keeping with the engineer's authority and allows the engineer to immediately react to a safety situation.

It is interesting to note that questions regarding the engineer's responsibility for site safety usually arise in the context of field resident services. Engineering services performed from the office may also involve site safety considerations. For instance, in the United States Court of Appeals for the First Circuit case entitled *Robert B. Reich, Secretary of Labor v. Simpson, Gumpertz & Heger, Inc. and Occupational Safety and Health Review Commission*, Case No. 922237, Simpson, Gumpertz & Heger (SGH) was cited for violating Occupational Safety and Health Administration (OSHA) standards when a SGH engineer told the general contractor during a telephone conversation from his office that certain metal floor decking would withstand a concrete pour. The pour proceeded and the deck failed, injuring five workers.

SGH appealed the citations arguing that OSHA's construction standards apply to physical work at a construction site, and not to a consulting engineer's professional services. An administrative law judge of the Occupational Safety and Health Review Commission agreed with SGH and dismissed the citations. The Secretary of Labor appealed to the First Circuit.

On August 20, 1993, the Court of Appeals affirmed the dismissal of the OSHA citations. Instead of relying on the argument that engineering services could not be construction work since such services do not involve substantial supervision of construction, the Court of Appeals found that the construction site was not a place of employment that SGH had a duty under OSHA to protect. The OSHA citations were therefore improper.

The SGH decision appears to shift the analysis from whether the engineer substantially controls the construction work to whether the engineer's place of employment includes the construction site. SGH did not have full time resident representation. Many project owners secure full time resident representation from the engineer. Whether such services cause the construction site to become a place of employment that the engineer has a duty to protect under OSHA remains a question.

5.13 The Hypothetical Revisited

Through poor business development practices, little to no personnel training and minimal quality control, the seasoned engineer has planted the seeds of dispute. The unprofessional manner in which the engineer has performed many of the services during construction has guaranteed a harvest.

The seasoned engineer's disregard for the accuracy of representations made at the pre-construction meeting demonstrates a disregard of professional practice. Misrepresentations will do little to establish a working relationship with the contractor and, furthermore, may be binding on the client to the client's disadvantage. A friendly pre-construction meeting is not going to avoid claims relating to poor engineering work.

The inaccuracy of the engineer's estimate at bid time is common and should present no real problem with a client who knows the limitations of an engineer's estimate. Yet our seasoned engineer makes the opposite point with the client and then magnifies the error by internally setting a higher standard for change order approval. Change orders should be measured against the objective standard of whether the subject work represents a change from the contract work, not whether the change order is going to be difficult to explain against tight project financing.

Fortunately, the project was awarded to a quality contractor who was apparently willing to work with the engineer. A good working relationship with the contractor, however, is not an excuse for exchanging favors. The seasoned engineer put his firm and client at risk by cutting a deal to avoid a delay claim with the groundwater problem. Furthermore, the seasoned engineer may have waived many of the claim-related contract provisions that he was hired to administer in cutting this deal. There is no question that certain accommodations reached between the engineer and contractor make sense and are necessary to promote smooth project performance. This accommodation, however, should never rise to a level that alters the contractual relationship between the contractor and client. By doing just that, the seasoned engineer has opened a Pandora's box of trouble.

The first trouble out of the box is a key substitution request that caused concern for even the green junior engineer. Yet, relying on the specification language (much of which the seasoned engineer ignored when it suited his purpose on the earlier delay claim), the seasoned engineer approves the substitution without proper review, thereby repaying the earlier favor to the contractor and committing clear malpractice.

It is no surprise that this poorly-designed project continues to deteriorate into a morass of change orders and claims. It is also no surprise that the substitution failed, thereby adding insult to injury. The seasoned engineer had failed to properly prepare the contract documents during design and then failed to properly administer those documents during construction.

In a last ditch effort to defend against the claims, the engineer raises the notice and waiver provisions of the contract. These same procedural provisions, long since ignored during construction, provide little defense at this late stage. The contractor will press the claims, project completion will be delayed and the contractor, owner, and engineer will end up in litigation.

Quite likely, the engineer, client, and contractor have failed to seek legal advice during project performance. Securing legal advice early on in a project, particularly a troublesome project as with the hypothetical, is in itself a dispute avoidance technique. Unfortunately, most engineers are penny wise and pound foolish, leaving legal counsel to pick up the pieces of failed projects after the fact and at great expense.

6.0 Partnering

The traditional relationship between the contractor and owner during construction is adversarial. The interest of the contractor in minimizing project performance to maximize profit is directly opposite to the interest of the owner in securing maximum performance at the least cost. Into this cauldron of competing interests steps the consulting engineer who is naturally aligned with the client/owner, yet called upon to administer the construction of his or her own design in a fair and unbiased manner. This is truly a no-win situation from inception and prone to disputes. Partnering addresses these competing interests by aligning the contractor's, owner's, and engineer's interests into a single project goal.

The U.S. Army Corps of Engineers pioneered the concept of partnering in seeking to resolve disputes associated with fixed-price low-bid public contracts. Essentially, partnering involves cooperative project management between the contractor, owner, and engineer. Post-award and pre-construction meetings between the principals of the project participants are held to develop mutual goals and strong working relationships early in the project. These mutual goals and relationships are reinforced through project performance by periodic team meetings.

By working together in what otherwise is an adverse relationship, the administrative burden of the project is streamlined and potential project problems are identified early on. The net result is fewer disputes.

Partnering only works if every project participant's willingness to work together is sustained throughout the duration of the project. Prompt and good faith efforts to avoid and resolve disputes are key to successful partnering. Defending one's position without compromise or taking the hard line on any dispute defeats the purpose of partnering.

III.

DISPUTE RESOLUTION

No matter how responsible the consulting engineer, certain disputes will arise that are simply unavoidable. The consulting engineer interacts with a number of parties, including owners, contractors, subcontractors, and suppliers. As previously discussed, all have competing interests bound by interdependent contractual rights and obligations. It is not uncommon for at least one party to take an uncompromising position that forces the remaining parties into more formalized means of dispute resolution.

The consulting engineer must first recognize the unavoidable dispute from those disputes that can either be avoided or quickly negotiated. Once identified as unavoidable, the engineer must take certain steps, the first of which involves insurance.

1.0 Insurance Considerations

Most consulting engineers carry errors and omissions insurance (E&O insurance) that characteristically charge ever–increasing premiums and deductibles. Such policies are essential to the practice of engineering and the protection of the consulting engineer's business. Typically, E&O insurance policies require that the engineer give notice of potential liability and to turn over the defense to the insurer. The insurer then assigns legal counsel of its choosing for the engineer.

Although the E&O insurer has the obligation to provide the consulting engineer with a defense, the insurer is in the business of making a profit under the policies. On a purely business level, it is in the insurer's best interest to pay out as little as possible on claims made against consulting engineers while incurring as small a defense cost as possible.

The engineer should not simply assume the matter will be taken care of by the insurer. The engineer has a vested interest in how any defense is handled for a number of reasons. First, given high deductibles, defense costs and claim payouts are in large part funded by the consulting engineer. Secondly, the insurer has little interest in the consulting engineer's business position *after* the dispute is settled. For instance, a dispute with a client involving a claim of malpractice may be best handled by an immediate and possibly generous settlement with the client. The consulting engineer may be able to salvage a long-term and profitable business relationship with that client as well as avoid tarnishing the engineer's reputation among other potential clients. The insurer's interest focuses on resolving the claim in the most cost-effective manner with secondary interest to the long-term business ramifications to the consulting engineer. The insurer, which controls the defense, may elect to vigorously defend against the client, thereby eliminating any chance of continuing a business relationship.

For this reason, the consulting engineer should take an active role in the defense of all claims, whether covered by insurance or not. The engineer may choose to seek separate legal counsel or request the insurer hire legal counsel of the engineer's choosing. At a minimum, the engineer should take an active role in the defense with whomever the insurer assigns to defend the matter.

2.0 Dispute Analysis

Whether the dispute involves only the consulting engineer and client or it involves multiple parties of the construction project, the positions of the various parties must be defined before resolution is possible. For instance, a contractor or subcontractor who believes that the engineer has caused compensable delay by failing to properly administer the construction contract cannot expect that such a claim will ever be resolved without a clear presentation of the basis and the supporting record for the claim. It is imperative that each party understand the basis that supports its own claim.

It is not uncommon for disgruntled engineers, contractors or owners to perceive some wrong, and demand that the wrong be corrected without spending the time in defining what went wrong. If an initial effort is made to determine what actions or omissions caused the problem, who was responsible for the actions or omissions, and how much in time and/or money was incurred because of such acts or omissions, the potential claimant may decide that responsibility is too unclear or that the amount of damages is too speculative to pursue the matter. In the alternative, if the claim is pursued, the potential claimant will be able to clearly define the basis of the dispute before proceeding.

The importance of clearly defining the dispute and potential recovery cannot be overemphasized. Dispute resolution is driven, not only by the "righteousness" of the claimant, but also by the cost in legal fees to vindicate that right. The cost in legal fees must be considered when evaluating potential recovery because legal fees will be a significant cost component. How does one go about evaluating legal fees? Consider the following:

1. Legal fees can run anywhere from $100 to $300 per hour.
2. Construction disputes involve complex and voluminous project records, all of which take time to review and understand.
3. Most lawyers do not have engineering or construction training resulting in additional time spent becoming versed in the field.
4. Similarly, a lawyer new to the field will be unable to separate the significant from the incidental project records, resulting in additional expenditures of time.
5. Few lawyers will take construction claim type disputes on a percentage of recovery (i.e., contingency fee) because the progress of the case and magnitude of work is so unpredictable.

It is not unusual for 30% or more of a claimant's recovery to be paid in legal fees. It is also not usual for a defendant to pay more in legal fees than what could have been paid to the claimant to settle the matter at the onset of the dispute. Whether a claim is worth pursuing or defending, cannot be determine without an estimate of legal costs.

Unfortunately, estimates of legal costs are notoriously inaccurate. The charges are not only driven by what the lawyer does to affirmatively represent your rights, but also what must be done in response to the other side's actions. Hence, most lawyers are very reluctant to provide any price estimate. Press the issue to secure at least a "ballpark" estimate. The "ballpark" estimate may clearly indicate that the claim or defense is simply not worth pursuing. Furthermore, if the claim or defense is worth pursuing, the ultimate legal costs will not be too much of a surprise.

3.0 Adversarial Dispute Resolution

The most common form of dispute resolution is adversarial in that each party takes adverse and bias positions and complies with certain agreed upon procedures in anticipation of the truth being revealed. Traditional adversarial dispute resolution includes litigation and arbitration, both of which are initiated with the filing of a claim.

3.1 Commencement of Adversarial Proceedings

Adversarial dispute resolution begins with the filing and service of a claim, which is referred to as a "demand" in arbitration and a "complaint" in litigation. Service is the term used for assuring delivery of the claim. This can occur in a number of ways, most notably, by sheriff, marshall, or process server.

A complaint details by numbered allegations what the defendant did that caused the plaintiff damages. The complaint identifies the theories of liability, or more simply, the reasons why the "plaintiff" should recover money from the "defendant." Finally, the complaint identifies the damages being claimed. Although complaints can ask for a variety of remedies other than money, the payment of money is, by far, the most frequent remedy. A demand in arbitration is similar to a complaint without the detail and is served by the "claimant" on the "respondent."

It should go without saying, that the service of a complaint or demand should not be handled by the mailroom or secretarial staff. There may be serious consequences to losing, ignoring, or failing to respond to such papers. Management must be informed immediately and, once informed, should refer the papers to counsel and notify the E&O insurer.

Once served with papers, do not call the plaintiff or claimant. If the project staff is initially informed of the ensuing litigation or arbitration, give clear instructions that the staff is not to contact the other party. Obviously, the project staff, deeply involved in the details of the project, may respond quite harshly to being informed that the firm is being sued over something they may have done. Whatever the response, it should be kept in-house and referred to legal counsel. Under no circumstances should the project staff contact the opposing party regarding the matter without advice of counsel. Communications between adverse parties before the various positions are defined is generally not helpful and, except in the context of settlement discussions, can be used against the engineer in later proceedings.

In response to the complaint, the defendant serves an answer. The answer responds to each of the allegations in the complaint by admitting, denying or claiming lack of knowledge as to the truth of each allegation. The answer also identifies defenses each party may have to the theories of liability presented and may include claims against the plaintiff, called counterclaims, claims against other defendants, called crossclaims, or claims against other parties, called third-party claims. All of these papers are referred to as pleadings.

An arbitration demand need not be answered since the liability claimed in the demand is considered denied in the absence of an answer. The respondent may elect to answer and assert a counterclaim as in litigation. The respondent may also elect to bring separate arbitration actions against other potentially responsible parties and seek consolidation of the proceedings.

Once all of the claims are asserted and responded to the "issues are joined" and the matter proceeds to the next stage of the process, which is discovery.

3.2 *Discovery*

Discovery is the term used to describe procedures that allow one party to investigate the substance of any other party's claims. In general, discovery includes document production requests that require access to all records except records of communications with counsel.

The engineer and staff must appreciate that whatever is put in writing may be disclosed. This is not to suggest that fewer project records should be kept, quite the contrary is true. What is suggested is that project records be kept as records of facts, not opinions. For instance, it is not unusual for a resident engineer to editorialize in a field report as to the contractor's ability to perform the work (e.g., "this guy couldn't build a doghouse") instead of keeping to the mundane facts of daily progress (e.g., "contractor did not show today"). It is always best to record the facts and keep opinions to oneself.

Discovery also includes interrogatories, which are a series of written questions that a party must answer in writing under oath. Interrogatories should not be taken lightly. Whoever is assigned to prepare draft answers to the interrogatories should do so comprehensively. There is an obligation to investigate the situation in order to provide a proper answer. If incomplete or evasive answers are provided, the Court may be asked to order proper and complete answers. This will cost more in legal fees than a proper answer would have cost in the first place.

Depositions, the most time-consuming of discovery methods, involves the taking of oral testimony from witnesses. At a deposition, the witness is sworn in and an attorney begins to ask questions. Construction litigation depositions of single witnesses can last for days, reflecting the complexity of most construction projects, magnitude of records involved, and tenacity of counsel.

Although discovery is less extensive in arbitration, it is very common for the parties to agree to an exchange of documents. It is less common for depositions to take place by agreement. The absence of depositions tends to cause less focused and more extensive examinations at the hearings.

The scope of a lawyer's inquiry at a deposition is extremely broad and it is not unusual for lawyers, unfamiliar with engineering or construction practice, to take the deposition inquiry into areas completely unrelated to the dispute at issue. This brings up a key point in the way engineers and lawyers are trained. Engineers as witnesses at depositions have the tendency to be helpful to the interrogating attorney. The answers given go beyond what is called for in the question, leading the interrogator into areas that would otherwise be missed. A deposition is an adversarial proceeding. The engineer, as a witness, is not present to educate the lawyer. Truthfully answer the question asked and say no more. For instance:

Q: Where is the project located?
A: Massachusetts.
Q: Are you licensed to practice professional engineering in Massachusetts?
A: Yes.
Q: Has your license ever been revoked?
A: Not in Massachusetts.
Q: Where has your license been revoked?
A: Maine, New Hampshire, Vermont, etc.

The proper answer to the third question was "No" as it would not have led to the next question.

Once discovery is completed, the case moves to trial, if litigated, and hearings, if arbitrated.

3.3 Litigation

Traditional litigation is premised in the theory that the presentation of adverse positions by lawyers to an unbiased jury will lead the jury to the truth. This theory, as it applies to the extremely complex and document intensive construction disputes projects is less than satisfactory.

Procedurally, litigation involves the presentation of the case by lawyers making opening statements, calling witnesses to testify, examination of those witnesses, cross-examination of opposing witnesses and closing arguments. The procedure and rules of evidence are designed to give each party a fair opportunity to present its side of the story. A jury (or judge in a "bench trial") then decides the truth.

Court practices and procedure originated in Colonial times and were designed to handle the types of disputes that the colonists encountered. As society became more complex, especially with respect to business transactions and relationships, the courts became less able to handle the more complex disputes. Construction litigation is a case in point.

Construction litigation is one of the most time-consuming, expensive, and inefficient means of dispute resolution. Courts generally shy away from construction litigation because of the burden in time it places on both the judicial system and, in the case of a jury trial, on the personal lives of each of the jurors.

Whether judge or jury, construction litigation at trial will require extensive testimony on engineering and construction practices. The judge or jury will not have this expert training and one of the lawyer's many responsibilities will be to educate the fact finder. This is no small task, given that an engineering or construction education takes years to attain, and now must be taught by a typically untrained lawyer to a completely unversed fact finder in record time. The best one can hope is that a very basic grasp of the complex issues is achieved.

Remarkably, a judge or jury's ignorance of the technical aspects of the field from which the dispute develops causes a greater dependence on the judge or jury's strongest asset, common sense. No matter how complex an issue, many disputes can be boiled down to the essentials and determined using common sense. It is this common sense that leads to vindication of the "right" party while arbitration may have lead to a more compromised decision. The risk with a non-expert judge or jury is that the lack of a firm understanding of the technical issues, regardless of common sense, will lead to an arbitrary or unfair finding.

Litigation, unlike arbitration, provides for appeal in the event of an arbitrary or unfair finding. Appellate courts are set up for this very purpose and the appealing party has the right to a careful review of any trial decision. Like everything, an appeal takes time and costs money. Also, the decision of the appeals court, like the trial itself, is public. In litigation, the parties cannot keep the dirty laundry of dispute resolution private.

Finally, the judicial system in most parts of the country is faced with overloaded staffs working within limited budgets. The courts are simply overwhelmed and unable to keep up with the ever-growing burden of litigation. The delays inherent in such an overwhelmed system promote frustration between the litigants and cause even more legal costs. Criminal matters must receive priority and civil disputes, like construction litigation, will receive secondary consideration. The net effect is that only financially well off litigants having expensive claims can afford to take advantage of full blown litigation and trial.

3.4 Arbitration

Arbitration involves a single or panel of arbitrators, expert in the field of construction, who decide the dispute after presentation of the case by the parties. The presentation involves statements made by the lawyers followed by testimony of witnesses called by the lawyers. The proceedings usually take place in a conference room and are informal and private. The most common rules of arbitration proceedings are defined by the Construction Industry Rules of the American Arbitration Association.

Arbitrators are selected by each party usually by way of rejecting unacceptable arbitrators from a suggested list until a single or panel of acceptable arbitrators remain. The arbitrators need not be lawyers, however, the absence of a lawyer from a panel of arbitrators may be somewhat problematic because the case will be presented by other lawyers. A lawyer panel member is more likely to exert control and authority over advocating lawyers than a non-lawyer arbitrator.

The key point is to agree upon a balanced arbitration panel, all expert in the field. If the dispute involves an engineer and contractor, for instance, an arbitration panel consisting of an engineer, contractor, and lawyer would be balanced and presumably address any inherent bias of individual arbitrators.

Arbitration is contractually based in that arbitration is only mandatory and binding if the parties have agreed so within a contract. If the parties have agreed to mandatory arbitration in the contract, they will be foreclosed from trying the matter in court, unless both parties agree to do so.

There are certain disadvantages to arbitration that must be noted. Multi-party arbitration, so common in construction disputes, faces a significant scheduling problem. Unlike a courtroom trial in which a judge sets a trial date and once the trial starts it will continue uninterrupted, arbitration proceedings usually go one or two days at a time and extend over, in some cases, years. This is because the parties, witnesses, lawyers and arbitrators must all coordinate their busy schedules in order to determine available dates. The arbitrators, usually have other employment responsibilities that cannot be ignored for extended and continuous periods of time. Hence, arbitration suffers from interrupted presentations causing significant delays.

Arbitration is not subject to the rules of evidence, and the arbitrators have complete discretion on hearing any evidence. The rules of evidence for courtroom trials prevent the introduction of certain testimony or documents as being unreliable. For example, the rules of evidence would cause hearsay to be excluded. In arbitration, any evidence whether reliable or not may be introduced. As a consequence, case presentations tend to take longer in arbitration than if litigated.

This situation is made more severe since one of the few reasons an arbitration decision can be appealed is if the arbitrators fail to hear all the proper evidence. As it is never clear what is properly excluded by the arbitrators, there is the understandable tendency to give the parties a wide latitude in presenting evidence. The net result is extended witness examinations and the introduction of voluminous construction project records. To the client, this means more hearing days and more legal costs.

With respect to the consulting engineering practice, arbitration offers one further complexity. Since arbitration arises from contract, it is possible that a single dispute involving multiple parties would require both arbitration and litigation to resolve unless all the parties have coordinated contract obligations requiring a single arbitration. Many consulting engineers refuse to include an obligation to arbitrate in their professional services agreement while including such a provision in the construction contract between owner and contractor. Under these circumstances, a contractor pursuing a claim against an owner in arbitration is less likely to independently pursue the engineer in litigation since the cost to do so would essentially double.

Similarly, the owner would be unable to "implead" the engineer into the arbitration with the contractor, something the owner would want do if the contractor's claim stems from the engineer's design or services. Like the contractor, the owner would be left with bringing a separate and expensive litigation against the engineer.

On the surface, this provides the engineer with limited protection in that no party wants to finance two adversarial dispute proceedings that could result in two inconsistent findings. In practice, however, the engineer's client, the owner, would not remain a client for long, once the owner discovers this potential difficulty in holding the engineer responsible for the engineer's own acts or omissions.

Arbitration suffers from a couple of common complaints that may well be more perceived than fact. For instance, arbitration awards are claimed to be a compromise between the various positions and rarely vindicate any party in total no matter how deserving. Arbitration is also thought to be more contractor- than engineer-oriented. There may be some truth in both of these observations depending on the panel of arbitrators. A well-balanced panel, however, will call the facts without compromise and bias.

Despite these disadvantages, the overriding benefit of arbitration is that the fact finder is expert in the construction industry. Theoretically, this will result in fairer decisions than those made by judges or juries unfamiliar with the field.

3.5 The Hypothetical Revisited

Recall that the substitute equipment has failed, the owner has refused to pay the contractor, the contractor has submitted a claim that has been rejected by the engineer for procedural defects, all culminating with the contractor walking away from the project. Neither the contractor nor engineer is taking responsibility for the substitute equipment failure and litigation follows.

Assuming the absence of an arbitration clause in the construction contract and professional services agreement, one litigation scenario is for the owner to sue both the contractor and engineer. The initial claims are outlined on Figure 18. The essence of the owner's claim against the contractor is that the substitution failed to meet the specification (breach of contract). The essence of the claim against the engineer is that, if the substitution met the specification, then

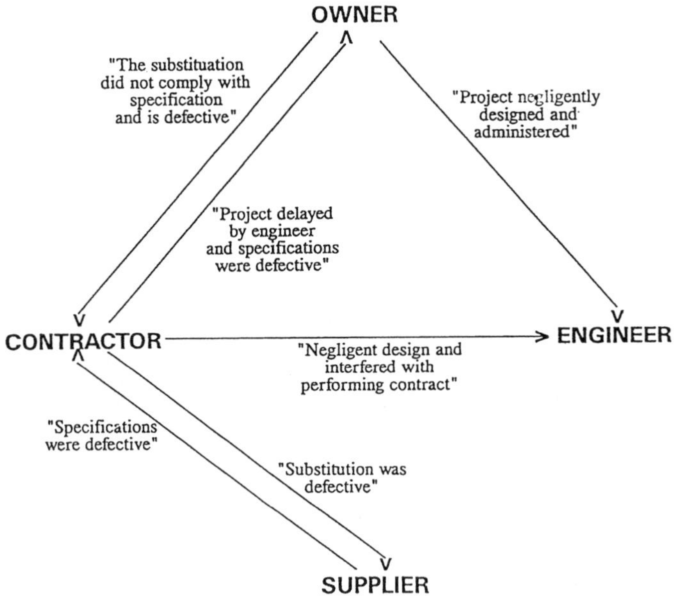

Figure 18.

the specification was improperly prepared, or in the alternative, the engineer improperly administered the project by allowing the contractor to supply a product that did not meet the specifications (negligence).

Note that the claims are internally inconsistent, in that if one is true, others must be false. Inconsistent allegations are a common practice, since the point of pleadings is to notify all parties as to *all* claims and defenses.

The contractor may then answer the owner's complaint by claiming the owner's specifications were defective and that the owner owes the contractor for its costs in attempting to meet the specifications including delay related damages caused by the engineer's administration of the defective specifications. A similar companion "cross claim" against the "co-defendant" engineer is also likely. The contractor may also claim against the supplier (referred to as "impleading" a third party) by taking the inconsistent position that, if the specifications were not defective, then the suppliers substitution was defective. The supplier will surely claim in response that, if its product failed, such failure was not due to product defect since the product met the specification requirements and was accepted by the engineer.

A mere casual look at Figure 18 demonstrates that the engineer is on the receiving end of most of the claims. This is not a promising position to be in, especially when discovery reveals what went on behind the scenes during performance of the engineer's design and construction administration services.

The engineer will be called upon to make some very embarrassing and damaging revelations. Certainly the project manager, design engineer and junior engineer will be deposed. Their combined testimony will establish a pattern of unprofessional conduct that will not only severely compromise the engineer's defenses but also tarnish the engineer's reputation. This pattern of conduct will be measured against the testimony of an expert engineer who has the luxury of Monday morning quarterbacking and will testify what should have been done and how the consulting engineer consistently failed to meet that standard of practice.

4.0 Negotiation

After discovery is complete and the majority of legal bills have been incurred, the parties are generally disenchanted, if not disgruntled, over the proceedings. Court or hearing dates are repeatedly continued which is a nicer way of saying delayed. Furthermore, when a hearing does take place, many judges make it clear that the Court is reluctant to tie up scarce judicial resources in trying a complex and time-consuming construction matter. Many judges actively encourage the parties to settle or seek some alternate form of dispute resolution. In the hierarchy of society's problems facing a judge on any given day, a complex construction dispute between business interests, is simply not going to be given top priority.

Hence, the parties to a construction dispute, burdened with extensive legal bills and facing further delays to a firm trial date that may be years after the original complaint was filed, realistically consider settlement.

It is one of the most unfortunate aspects of adversarial dispute resolution that productive settlement negotiations rarely take place before extensive discovery has taken place. Early on the parties and counsel are usually more interested in posturing than resolving. Some discovery is usually necessary to clarify claims. Most cases, however, are subject to unnecessarily extensive discovery that rarely leads to unexpected revelations but always increases legal costs. The legal costs of discovery usually has the tendency to "soften up" the parties' initial positions leading to more fruitful settlement negotiations. By the time discovery is near completion, the legal process has taken its toll and the parties no longer insist on principle, instead focusing on the business realities of getting out of the matter without incurring further expenditures in time or money. In short, no party is looking to win, all parties are looking to minimize the damage. What may have been the absolute bottom line early on in the process now becomes negotiable.

It is somewhat surprising that lawyers can be obstacles to settlement at this stage. Trained to advocate, the lawyer may well have been advising the client as to the righteousness of a certain position and zealously pressing that position during what may be years of protracted litigation or arbitration. The lawyer takes on the position as his or her own cause, unfortunately at the client's expense.

Although most of the important law in this country stems from personal sacrifice made by lawyers taking on a client's cause, construction litigation or arbitration rarely, if ever, addresses those societal or constitutional issues that justify such unbending sacrifice. Construction litigation or arbitration is first and foremost a business concern to the engineer, contractor and owner. As such, the client's interest is to resolve the matter as fairly and as expeditiously as possible all in keeping with that business interest. To that end, lawyers should encourage settlement negotiations.

It takes the cooperation of all parties to reach settlement and, conversely, one party can nix any potential settlement between the others quite easily. Most construction disputes involve multiple parties. The more parties, the more likely one party will insist on having a "day in court." Hence, it is not uncommon for the parties to need help in negotiating a settlement. The help is available with Alternative Dispute Resolution techniques.

5.0 Alternative Dispute Resolution

Given the disadvantage of arbitration and litigation, alternative forums of dispute resolution (ADR) have and are being offered. ADR is limited only to the ingenuity of the parties in tailoring a mechanism that assists in reaching dispute resolution. In its broadest form, ADR is a procedure proposed by and agreed on between the parties as an alternative to traditional litigation or arbitration. The procedure can be as formal and as binding as the parties desire. ADR can take place at any time. ADR can be used as an alternative to, or in conjunction with, ongoing litigation or arbitration. ADR can be designed to make negotiations productive or as a vehicle in reaching a binding resolution. Common forms of ADR include mediation, minitrial, step negotiation, standing neutral, and escrow bid documents.

The relative costs of ADR as against litigation and arbitration is depicted on Figure 19. ADR procedures are cost-effective and are much less adversarial than

Relative Costs of ADR Methods

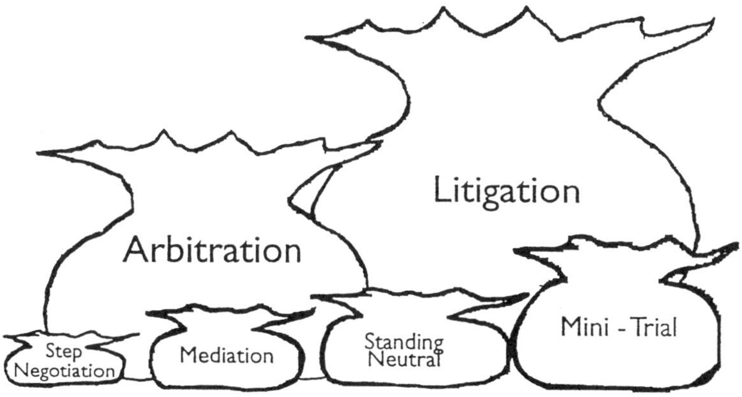

Figure 19

litigation or arbitration. The less adversarial the proceeding, the more likely the parties can salvage business relationships that would otherwise be lost.

5.1 Mediation

Mediation is formalized negotiations during which a mediator facilitates a settlement. Each party is generally given the opportunity to present its position. The mediator then confers with each party individually and encourages each party to honestly evaluate the strengths and weaknesses of its position.

After consulting with each party individually, the mediator will have developed a good sense as to which party has taken the most unreasonable posture. Generally, the mediator will focus on the more unreasonable party and encourage compromise.

Once the mediator knows how distant the parties are from settlement, the mediator can determine whether continued negotiations would be fruitful. If it is clear that the parties are within settlement range, the mediator can then use his or her skills in causing each party to incrementally compromise to close the gap between what one party is willing to pay and the other is willing to accept. At this point, each of the parties has a fairly strong incentive to settle simply because the gap has already closed considerably and the alternative is more legal expenses and the risk of trial or arbitration.

Mediation is entirely voluntary. Any party can simply refuse to mediate and stop the proceedings. Most parties who have agreed to mediate, however, have demonstrated a willingness to compromise and have a vested interest in making the mediation successful. In addition to a good faith interest in settlement, the key to successful mediation is in the skill of the mediator. The American Arbitration Association as well as many courts maintain lists of skilled mediators. The engineer is well advised to take advantage of that list.

There are few good reasons not to participate in mediation. One of the more common claimed disadvantages of mediation is that the mediation process discloses a party's trial posture. The theory goes that by keeping one's case to oneself allows that case to be more effectively presented at trial.

With liberal discovery available, it is a rare claim or defense indeed that goes undetected to the eve of trial. For this reason, there is little merit in avoiding open and good faith mediation. In fact, there is advantage in presenting the strongest case possible in mediation as a means of encouraging the other party to compromise.

5.2 *Minitrial*

Private ADR services generally offer the services of an experienced dispute resolver, such as a retired judge, to hear and ultimately decide the issues presented. The parties agree to split the cost of the services and abide by the decision rendered. The proceedings are tailored to meet the needs of the parties and promote timely resolution. The proceedings can be as formal as trial practice or as informal as summary presentation of positions.

Essentially, minitrials are private court proceedings at a fee. These proceedings have all the advantages of civil litigation without the inefficiencies and delays of the public court system. Many parties resent paying a fee for what is a citizen's right to a day in court. Given the overwhelmed and undersupported public system, the wait for that day in public court may be more expensive than splitting the fee for a timely day in private court.

Minitrials can be tailored to address the entire dispute or limited to resolve certain issues that are effectively stalling productive negotiations and settlement. Similarly, the minitrial's decision can be made binding (as in arbitration) or advisory as an aid to subsequent mediation or negotiations. As with all ADR, the procedure and jurisdiction of the process is tailored to the nature of the dispute by agreement of the parties.

For example, one dispute resolution procedure that encourages the parties to take reasonable positions towards settlement is baseball arbitration. Baseball arbitration is modeled after arbitration proceedings that have found application in resolving salary disputes of major league ball players. As applied to construction disputes, baseball arbitration requires the parties to submit their most reasonable offers to a third party after presentation of the case in summary fashion. The neutral third party, who will likely understand the issues after summary presentation as well as, if not better than, a judge or jury after a full trial, then decides the matter by accepting one or the other offer as the fairest. There is no compromise or splitting the difference. If the respective offers are reasonably close, the parties are offered the opportunity to settle the difference prior to a binding decision.

Under these circumstances, each party has a great incentive to be as reasonable as possible since to be otherwise risks losing the matter to the other party's most reasonable offer.

5.3 Step Negotiation

Step negotiation is a formalized procedure calling for the systematic review and determination of a dispute from field staff through project management staff. Step negotiation is contract based in that the parties prospectively agree to process disputes through step negotiation prior to adversarial resolution. For example, the construction contract may require each dispute to first be brought to the field resident's attention. The field resident then prepares a recommendation that, if accepted, resolves the matter and, if rejected, is referred to the next higher level of authority within the project team, such as the project engineer. The project engineer makes a recommendation which again, if rejected, is referred to the next higher level, such as the project manager. This process of review, recommendation, and appeal, all required prior to filing any legal action, eventually reaches project management.

Under a step negotiation program, which for all practical purposes formalizes what takes place on most projects in any case, each level of authority has an incentive to reach resolution to avoid involvement of the next higher level. Hence, the field resident wants to resolve the dispute at the field level. Similarly, the party pressing the dispute, recognizes that most project engineers and managers will stand by a field resident's recommendation in the first instance. Hence, this party also wants a field resolution. As a consequence, the success of step negotiation resolution, like the success of many aspects of the project, may rest with the competence of the engineer's field resident and the contractor's superintendent.

5.4 Standing Neutral (DRB)

The standing neutral process involves the referral of a dispute for resolution to a neutral third party. Like step negotiation, the standing neutral concept is agreed to by the parties in the construction contract. During the life of the project, the parties support a third party neutral to whom disputes are referred as they develop.

The most common neutral third party is a dispute resolution board, which, in most cases, has the authority to make non-binding recommendations for settlement. The recommendations may then be admissible in a later legal proceeding. The compensation of the board, procedures for claim presentation, decision-making authority of the board, rights to appeal, and finality of the board's decision are all detailed by the construction contract.

The dispute resolution board is generally balanced with an owner-sponsored and contractor-sponsored member. A third member, usually the chairperson, may be selected by the other two members and approved by the contractor and owner. Once in place, the DRB holds hearings that may combine elements of arbitration, minitrials, and step negotiations. For instance, the DRB procedure for Boston's Central Artery Project, depicted on Figure 20, illustrates how a dispute is first determined by the owner's representative, appealed to the DRB, and then appealed to the project director before litigation can take place. At any one of these steps, the determination can be accepted and dispute resolved. Furthermore, failure to reject a determination results in that determination becoming binding, thereby resolving the dispute. This concept promotes real-time dispute resolution.

DRB's have proven quite effective. An excellent study of DRB experience is presented in the ASCE's 1991 publication entitled *Avoiding and Resolving Disputes During Construction* by the Technical Committee on Contract Practices of the Underground Technology Research Council to which the reader is referred.

5.5 Escrow Bid Documents

Escrow bid documents is a dispute resolution aid. Under an escrow bid document requirement, bidding contractors supply certified bid estimates with the bids. The bid estimate of the successful contractor is then held in escrow, and those estimates of unsuccessful bidders are returned. When a dispute arises, the bid documents serve the limited purpose of assisting the engineer, owner, and contractor in reaching a resolution as to damages. The contract documents may require disclosure of the escrow bid documents as part of more formalized alternative dispute resolution procedures.

Escrow bid documents resolve any question as to what the contractor carried in its bid. If this question is not central to the dispute, such documents are of little value. Escrow bid documents will not assist in resolution of disputes unrelated to the contractor's interpretation or pricing of the original as-bid contract.

5.6 The Hypothetical Revisited

The seasoned engineer may be well advised to make more than a generous offer of settlement during negotiations in order to avoid a trial that could both tarnish the engineer's reputation and bury the engineer in an adverse finding.

If negotiations falter, the engineer should encourage alternative dispute reso-

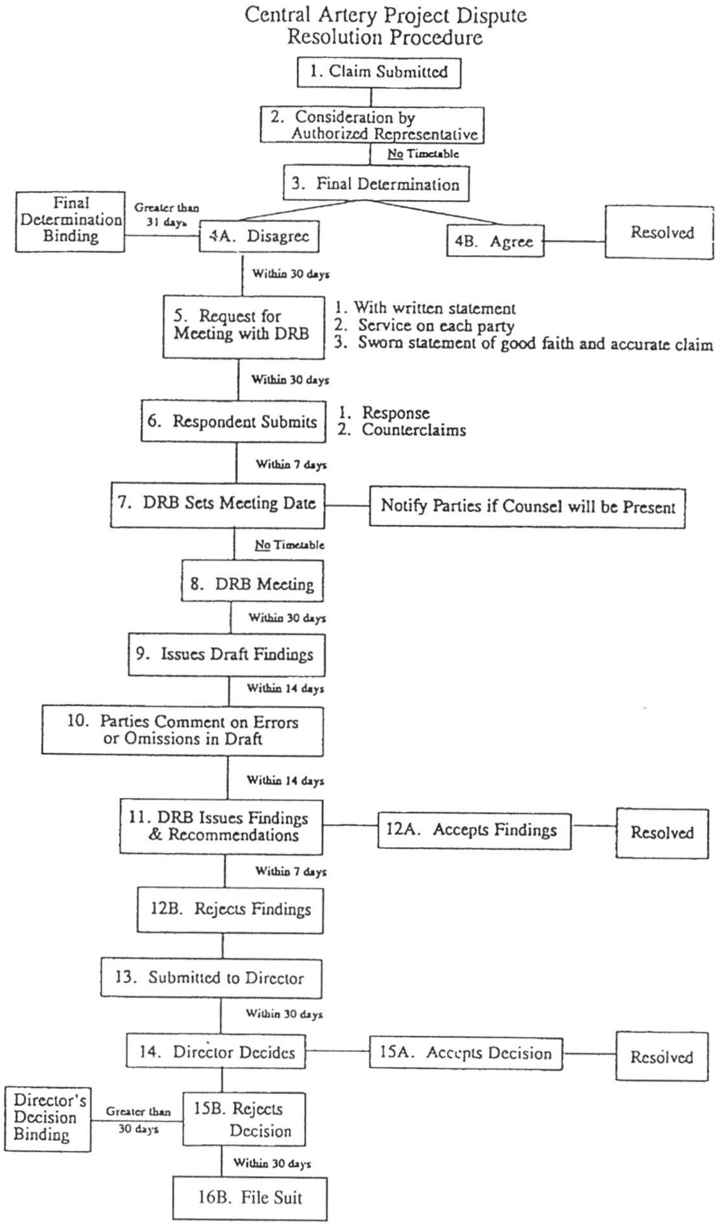

Figure 20.

lution that will allow for a private settlement of the claims. The engineer's insurer should be made aware of the business ramifications of drawn out litigation and public trial.

In the hypothetical, the engineer's behavior was clearly unprofessional. The outcome will prove much more costly to the engineer than the cost of responsibly practicing engineering and avoiding disputes in the first place.

IV.

Conclusion

The traditional design and construction administration services provided on both public and private construction projects are, to say the least, dispute prone. No party to the project, except possibly a very naive owner, begins construction without anticipating the inevitable problem. As we have all learned from Murphy's Law, something always goes wrong and usually at the worst time.

Disputes, however, are not necessary consequences of Murphy's Law. Disputes, instead, are when the parties to a project cannot agree on how to fix, or more likely, who is to blame for the inevitable machinations of Murphy's Law. Disputes then require effort on the part of the parties to blossom and flourish. Many disputes are well nurtured and cared for in construction projects. Some, in fact, take more energy to tend than the original problem would have taken to address. Notwithstanding this wasted effort, the parties to a construction project let disputes take on lives of their own, and in some cases, the disputes outlive the project itself.

The best dispute is the avoided dispute. Engineers can avoid most disputes by simply practicing the profession in a competent and responsible manner. However, no matter how competent or responsible, a few disputes cannot be avoided.

For these unavoidable disputes, the engineer should consider realistic settlement that focuses on the business realities of the situation. Seek compromise and timely resolution. In the absence of compromise, seek agreement on the method to resolve the dispute. Consider alternative techniques that bring about timely, and no less fair, resolutions.

Keep in mind that protracted adversarial dispute resolution is an expensive endeavor rarely resulting in vindication. Even when vindication is realized, it comes at a significant cost. The winner in adversarial dispute resolution is usually the party who walks away losing the least.

V.

Suggested Reading

1.0 Engineering Practice

1. "Quality in the Constructed Project—A Guide for Owners, Designers and Constructors," (1990). *ASCE Manual and Reports on Engineering Practice No. 73*, Volume 1, ASCE, New York, N.Y.
2. *Standard Form of Agreements.* Engineers Joint Contract Documents Committee—National Society of Professional Engineers, American Consulting Engineers Council, American Society of Civil Engineers, New York, N.Y.
3. *Peer Review Program Manual.* (1985). American Consulting Engineers Council, Washington, D.C.
4. Avoiding and Resolving Disputes During Construction. (1991). Technical Committee on Contracting Practices of the Underground Technology Research Council of the American Society of Civil Engineers, New York, N.Y.

The American Consulting Engineers Council and American Society of Civil Engineers have numerous publications and are an excellent source on engineering practice.

2.0 Legal Considerations

1. Walker, N., Walker, E., and Rohdenburg, T. (1979). *Legal Pitfalls in Architecture, Engineering and Building Construction*, McGraw-Hill Book Company, New York, N.Y.
2. Stokes, M. (1977). *Construction Law In Contractor's Language*, McGraw-Hill Book Company, New York, N.Y.
3. Dunham, C., Young R., and Bockrath, J. (1979). *Contracts, Specifications and Law for Engineers*, McGraw-Hill Book Company, New York, N.Y.
4. Sweet, J. (1977). *Legal Aspects of Architecture, Engineering and the Construction Process*, West Publishing, St. Paul, Minn.
5. Cushman, R. and Bottum, T. (1987). *Architect and Engineer Liability: Claims Against Design Professionals.* John Wiley & Sons, Inc., New York, N.Y.
6. Acret, J. (1984). *Architects and Engineers Their Professional Responsibilities*, Shepard's/McGraw-Hill, Colorado Springs, Colo.
7. Cushman, R., Carter J., and Silverman, A. (1986). *Construction Litigation: Representing the Contractor*, John Wiley & Sons, New York, N.Y.

Index